# THE CROWD AND THE COSMOS

# Praise for *The Crowd and the Cosmos*

'*The Crowd and the Cosmos* is a superbly written insight into the unique and powerful contribution enthusiasts from all walks of life can make to scientific knowledge. It is also a fascinating and much-needed description of how we acquire reliable knowledge about Nature, from the search for planets and perhaps civilisations around distant stars to observations of Penguins in the Antarctic and what they can teach us about the impact we are having on our own world.'

Brian Cox

'Chris Lintott, is a modest genius. He has quietly revolutionised modern Astronomy (and a few other branches of science) by using digital platforms to involve the public in processing data. Essentially anyone who wants to contribute some of their spare time can, and is invited through Chris's Zoooniverse projects to do real science. Literally millions have taken up the invitation. This is a beautifully readable book, which tells the story of the Zooniverse and much more. Chris is delightfully anecdotal, inclusive and witty, yet never shirks in-depth explanations of the cutting edge science he's delivering to us, almost before we realise it! This is the New Age of Science for All!!!'

Brian May

'*The Crowd and the Cosmos* gives an authentic flavour of astronomical research and its appeal. But it's especially significant because it offers a first-hand account of how Chris Lintott conceived and led the "Zooniverse" project, thereby enabling huge numbers to participate in significant research, and even make important discoveries. His pioneering initiative has spawned similar programmes in naval history, conservation, and other subjects—triggering a benign social revolution in scholarship and education.'

Martin Rees

CHRIS LINTOTT

# THE CROWD & THE COSMOS

## ADVENTURES IN THE ZOONIVERSE

OXFORD
UNIVERSITY PRESS

OXFORD
UNIVERSITY PRESS

Great Clarendon Street, Oxford, OX2 6DP,
United Kingdom

Oxford University Press is a department of the University of Oxford.
It furthers the University's objective of excellence in research, scholarship,
and education by publishing worldwide. Oxford is a registered trade mark of
Oxford University Press in the UK and in certain other countries



First Edition published in 2019
Impression: 1



Published in the United States of America by Oxford University Press
198 Madison Avenue, New York, NY 10016, United States of America

British Library Cataloguing in Publication Data
Data available

Library of Congress Control Number: 2018967074

ISBN 978–0–19–884222–4

Printed and bound in Great Britain by
Clays Ltd, Elcograf S.p.A.

# CONTENTS

# PREFACE

There is a faint star, dim and red, which shines feebly in the constellation of Aquarius. You would need a decent telescope to see it at all, and in all of humanity's history of studying the Universe no one bothered giving it a name. The star, recorded in catalogues as J23154776-1050590, is about 600 light years from Earth; not too far on cosmic scales, but close enough that the light we see now set off in the early fifteenth century, when King Henry V's English army was fighting the French at Agincourt.

Andrew Grey is a car mechanic who lives in Darwin, in northern Australia. He's also an amateur astronomer with a collection of telescopes used for stargazing under the clear desert skies, and in April 2016 he was the first person to realize that J23154776-1050590 was a star worth keeping an eye on. Every so often, the star does something odd. It winks, dimming slightly for an hour or two.

These winks reveal the presence of something that would otherwise be hidden. They're caused by the regular passage of a family of planets that happens to cross (or 'transit') the face of the star as seen from Earth. The effect is subtle—planets are small compared to stars, and a single wink results in a dimming of much less than 1 per cent of a star's brightness—but we can see them, and the immutable laws of physics dictate that as a planet completes orbit after regular orbit around the star, that single dip will be followed by another exactly one orbit later, and another and then another, each adding to observers confidence that the planet really exists.

Thanks to Andrew's ability to notice these small changes in brightness, we now know that J23154776-1050590 has at least five

**Figure 1** Artist's impression of five worlds around K2-138, discovered by citizen scientists in the Exoplanet Explorers project.

planets in its system. They crowd around the star, now given a shorter catalogue number: K2-138 (Figure 1). Nor were these new planets just another entry in the rapidly growing planet catalogue. Each is closer to their star than Mercury is to the Sun. Packed in tightly, they form a resonant and harmonious pattern, each world completing nearly exactly three orbits in the time taken for the next one out to go around twice, an arrangement which might persist for billions of years and which contains within it secrets about these world's formation.*

For centuries, and perhaps longer, astronomers dreamt of discoveries like this. Yet finding new planets, and new solar systems, is now something that you can do at home. Andrew's discovery was made not with one of his telescopes, but with a web browser. The star was one of many monitored by the *Kepler* space telescope, launched and operated by NASA, whose team gave the

* While editing the book, we were able announce the discovery of a sixth planet which nearly fits the pattern, but is slightly off. Even more intriguing!

data it collected away for free. A team of astronomers at Caltech had looked for likely planets, and shared their analysis online via a website that allowed anyone to show up and help. A discovery that twenty years ago would have put you in line for the most prestigious prizes in science is now something that you, the reader of this book, might manage in an otherwise idle lunch hour.

Access to data from telescopes all over the world is now something that astronomers take for granted and, as it turns out that being open with each other means that we're also open to the world, an increasing number of people are joining us in exploring the Universe. Consider Despina, an obscure moon of Neptune which was first discovered in images taken by the *Voyager 2* space probe as it approached the ice giant in the summer of 1989. No other probe has passed this way, and so we know very little about Despina, other than the fact that it's small, just 150 kilometres across, and that it lives just inside one of the planet's dark rings. It is close enough to Neptune, in fact, that it is probably spiralling slowly inwards, compelled by the push and pull of tides induced upon it by the planet's gravity.

One day it may fall apart completely, but for now, there it sits. *Voyager 2* shot through the Neptunian system at high speed, and observations of the planet itself and the largest moon, Triton, were the priority. Beyond noting its existence, little was done with Despina in the short interval between discovery and fly-past, and so all we have had for the last thirty years are a small set of images that make it look like a speckled jelly bean. *Voyager*'s encounter with Neptune was one of the things that made me, as a schoolkid, avidly interested in space, but I can't say that the diminutive moon made much of a mark.

Despina is part of this story because of some amazing detective work done by professional philosopher and amateur astronomer

Ted Stryk. Since *Voyager* flew past, the best images of both Uranus and Neptune have come from the *Hubble Space Telescope*, and in 2006 Ted saw a *Hubble* image which showed one of Uranus's larger moons, Ariel, transiting the disc. It's a fun image, with both Ariel and its shadow sharp against the pale green disc of the planet, and Ted wondered if there were any similar shots from *Voyager*.

I would have bet good money on the idea that there was nothing new to find in the *Voyager* dataset. Some missions produce enormous libraries of images, with plenty to keep scientists and their friends going for decades. The *Voyager* probes, though, especially when in the outer reaches of the solar system, relied on a fairly low bandwidth antenna to get data back to Earth, and so relatively few pictures were ever sent. Each image from this expensive, once-in-a-lifetime mission, had surely already been extensively studied.

Ted, though, found something new. He got hold of the raw data, and used modern technology and his expertise in image processing to see something no one else had. In a sequence of images taken over the course of nine minutes on 24 August 1989, a small black dot can be seen on the planet's blue, cloud-streaked face. In one of the images, and only one, a small bright dot appears near the edge of Neptune's disc.

This, remarkably, is Despina, caught in transit during the Neptune encounter as seen by the speeding *Voyager* probe (Figure 2). The first dot, which appears in the whole sequence, is the moon's shadow, and the second dot the moon itself, just entering the disc. It's a beautiful and poetic set of images, a moment in time captured during the only visit of a human-built craft to the most distant planet in the Solar System—and it may be useful too. Despina's orbit isn't well known, and pinning down its presence in this particular set of images will help work out how it behaves, and what its ultimate fate will be.

**Figure 2** Despina, seen as a bright dot accompanied by its moving shadow as seen by *Voyager* 2 and discovered by Ted Stryk. This is a montage of four images, taken nine minutes apart.

Despina isn't an isolated example. Planetary scientists on missions to the planets now regularly collaborate with a loose network of image-processing experts to get the best out of their data. When two Mars rovers, *Spirit* and *Opportunity*, landed on the red planet in 2004, the team behind them, led by Steve Squyres and Jim Bell, made the decision to make the data their robot explorers sent back available to the public as soon as it was received by NASA. If you happened to refresh the web page at just the right time, you could see an image taken on the surface of Mars before anyone else on Earth.

The images have allowed scientists to show that what is now a rusty desert was once a wet world. Mars, it's now clear, once had rivers and seas and lakes and oceans. The images from the rovers have also fuelled the dreams and fired the imaginations of a community of fans back here on Earth, many of whom collaborated to make use of the images sent back. This community had time (and often, the skills) to do what the scientists could not, making colour versions of images, charting their journeys on elaborate maps, and creating mosaic views of landscapes and the odd rover selfie. When the venerable magazine *Aviation Week* wanted to put *Spirit* on their cover, the image they used was created by four outsiders, who collaborated online without ever meeting. (One of them, Doug Ellison, a graphic artist from Leicester, now works at the Jet Propulsion Laboratory and was part of the team that operated the rovers.)

Whether it's discovering new worlds or exploring Mars, the web allows each of us to be part of the scientific enterprise. It's not just planets either. A distributed network of volunteers have spent the last decade sorting through images of galaxies, and mapping our own Milky Way. Others have helped conservationists and scientists study and monitor animals, ranging from lions in the Serengeti to coyotes in Chicago (and, lest you think this is about exciting, charismatic species, they have also spent time looking at many blurry images of kelp off the Californian coast and watched videos of egg-laying intestinal worms). Old documents, from ancient Greek papyri to the records and letters of anti-slavery campaigners in nineteenth-century America, have been explored and transcribed by still more volunteers. Hundreds of thousands of people have taken part in projects like this; in this small corner of the world wide web, together they have contributed to our understanding of the world and the cosmos.

I find it inspiring, and at a time when we tend to talk about the internet and the communications revolution it has precipitated in mostly negative terms, it's a reassuring reminder that the vast majority of people, both individually and collectively, are good. Even when assembled as that most modern of bugbears—a crowd on the internet—they are capable of remarkable feats of both generosity and skill. I hope reading this book will inspire many people to rush to their nearest screen to try to find a planet for themselves—data from NASA's new planet hunter, the *TESS* satellite, will be flowing in torrents by the time you read this book—or, more simply, to take a more personal interest in the Universe.

This book is the story of how I, a distractible astronomer, ended up watching all of this activity unfold from a grandstand seat. Each of the examples I've mentioned so far are from projects that live on a platform called the Zooniverse, built by a mercurial and talented team of web developers, scientists, and educators which I've been proud to lead. I haven't always had space to stop and explain who did what, or how a million conversations led us collectively to solve the problems whose results are presented here, but you should be aware as you read that everything we've been able to achieve in the last few years is the result of work by a team full of people much smarter than I.

It's very easy to forget when describing a project such as the Zooniverse that the technical approach taken is at least as important as the science, and in this case everything we've done has been shaped by early decisions to take both halves equally seriously. That we did this was due to Arfon Smith, who was my co-conspirator and the technical lead for the Zooniverse's crucial early years, and I would be remiss if I didn't thank him here for that insight and all the hard work. Among many others Lucy Fortson and Laura Trouille in particular also deserve my gratitude and thanks for their leadership and support. The Galaxy Zoo team—particularly Karen Masters and Bill Keel—have taught me an enormous amount, and been very tolerant of my distractions.

In these pages, I've focused mostly on our astronomical projects—they're what I know best, after all, and they say a lot about what modern astronomy has taught us about the Universe, and what we still don't know. My bias as a scientist is the same as it was years ago when I was a schoolboy reading about Neptune and *Voyager*—I care about space most of all—but a lot of the thrill of the Zooniverse has been the license to get interested in other people's research worlds and to draw ideas and inspiration from scholars of the humanities, climatologists, zoologists, and more. This book in particular was being written during the time the Zooniverse team was participating in the Connecting Scientific Communities project, funded by the Arts and Humanities Research Council and led by Sally Shuttleworth, and it is a very different beast because of the conversations we had.

As that implies, the book you're now holding is very different from the one I set out to write more than five years ago and many people have had a role in getting it into your hands. As well as support from many close friends, I want to mention Pedro Ferreira—who was finishing his brilliant book on the history of gravity, *The Perfect Theory*, as I was starting—who was a huge encouragement, and Rebecca Carter, my long-suffering agent, deserves all my thanks too, along with the team at OUP led by editor Latha Menon. Expert reviews of parts of the text were provided by Katharine Anderson, Chris Scott, and Brooke Simmons; any errors that remain are, of course, still mine.

I've tried to write in a way that conveys how much fun I've had in the Zooniverse for the last ten years, enjoyment that's due almost entirely to the efforts of our incredible community of volunteers. A small number of names is included at the back of this book, but I wish I could include them all. Their desire to get stuck into almost any problem is continually a source of inspiration and wonder, and it's been a pleasure working for them. If you're one of them, even if you only participated for a few minutes, thank you for what you've helped us all learn. If you haven't yet dived in, then I hope that this book inspires you to become an active participant in the grandest of all adventures—our attempt to understand the Universe, and our place in it. See you in the Zooniverse.

Chris Lintott,
Oxford, November 2018

# 1

# HOW SCIENCE IS DONE

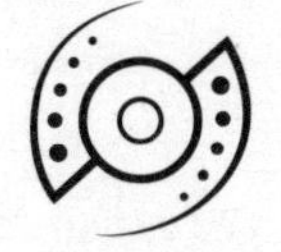

What does science look like? Is it a blackboard covered with confounding equations, a set of seemingly mystical and obscure symbols inscribed in chalk by a tweedy professor? Is it a laboratory filled with bubbling chemicals, or an expedition deep into the Amazonian jungle? Maybe it's a set of staccato sentences, delivered in front of dramatic backdrops by a suspiciously enthusiastic television presenter over a soaring soundtrack, or maybe it's just the mysterious set of knowledge that means I can have a new iPhone but which ensures also that its battery life will be measured in minutes. It's why dropped things in the kitchen will head for the floor, and also why toast lands buttered side down.*

It is all of these things, but to most people the need for science to speak the language of mathematics, the associated rigour, and a perception that to dedicate oneself to science means an unswerving devotion to the passionless weighing of competing hypotheses adds up to a vision of a grand but cold and impersonal edifice. Science, whether encountered out in the wild or in

* Not as random an example as you might think! See Matthews, R. A. J., 1995, Tumbling toast, Murphy's Law and the fundamental constants, *European Journal of Physics*, 16, 4: 172–6.

a battered school textbook, seems established in ground far removed from normal human concerns, more of a secret lab in the desert than part of our everyday human lives.

It's this perception that creates the stereotype of a scientist as being outside normal culture, the high priest of a technocratic caste, a group with their own language and concerns. Sometimes, this perception can be flattering—a typical response when I tell someone I'm an astronomer is for them to assume that I must be 'smart'—but those of us who spend our time on this thing called science know the reality is very different. Our scientific research is as much part of the real world as last night's takeaway. Science is—it has to be—a human pursuit. When progress comes, it arrives not out of the blue, but as the result of hopes and dreams, followed by blood, sweat, and not infrequently the tears of normal human beings. What's more, this is as true of the works by Newton, Darwin, or Einstein that we celebrate as the great solo masterpieces of the genre as it is of the great collaborative projects like the Large Hadron Collider (LHC) or the Human Genome Project which bring together thousands of people from hundreds of institutions to produce science on almost industrial scales. It's easy to forget when reading about the latest medical advance in the newspaper, or when listening awestruck to an astronomical discovery that requires liberal use of the word 'billions', but each advance in knowledge is won because someone out there wanted it badly. Scientific truths don't drop from the sky; they are worked for and fought over.

Knowledge expands because of the effort we put in, and the results can inspire. In the past few years we have come to know for certain that when we look at the night sky we are seeing stars that have planets in orbit around them, just as our own Earth and its companions circle the Sun. Just knowing that fact really does change how I look up at the heavens. I find it hard to think about without being impressed at our species' cleverness, and at our ability to figure things out. I feel the shock, and the awe, of being part of a species capable—perhaps uniquely so—of understanding our place in the Universe.

The enormous changes wrought by digital technologies have, as I've already mentioned, made it possible for everyone to take part in that effort, perhaps for the first time in human history. Whether you want to classify rare beetles at the bottom of the garden, or be the first to explore part of the Martian arctic in satellite images, it's clear that we can no longer indulge in the twentieth-century habit of leaving science to the scientists, but in area after area we are finding that we must instead all pitch in.

I believe that finding places where we can all make contributions to science is good for the progress of research, accelerating the pace of discovery and preserving all sorts of options that would otherwise be closed off, but it is, I think, also good for each of us to find a little time to make a meaningful contribution to our understanding of the world. Many of us need a new relationship with science, one based on mutual respect and not only on listening to the reporting of impressive feats of derring-do.

It's become a cliché in writing or talking about science communication to conjure up a 'typical' dinner party, usually in North London for some reason.* Conversation has somewhat inexplicably turned from house prices and schooling (the only really acceptable topics for imaginary North London dinner parties) to something sciency. Maybe it's the recent arrival of the European Space Agency's Rosetta probe at Comet Churyumov–Gerasimenko, whose pictures made the front pages of papers, or

* I think there's been so much written about this stereotypical dinner party that we should doubt whether anything like it ever occurs. If anyone has actually been to a dinner party in North London and tried to discuss science, only to fail, then do get in touch. I'll buy you a mid-price Portuguese wine as a reward.

maybe it's the most recent appearance of an old perennial news-friendly headline like the 'discovery' that red wine either causes or cures cancer. The details don't matter; driven to extremes by such a swerve in conversation, so the story goes, someone will quickly volunteer that they never understand science, that it left them cold at school and they could no more distinguish an asteroid from an adenoid than design a rocket and fly to the Moon.

The point is, I suppose, that at such an imaginary dinner party it's much harder to believe that anyone would say that they don't read for fun, or that they never understand cinema, or politics. Science can be dismissed without shame, and this says something about its status in our society. We could, and probably should, make the same point by noticing that a claim to scientific expertise is often followed by an admiring exclamation of assumed authority, but both are essentially expressions of fear, a sense that science is something 'other' to be pursued only by specialists.

Pursued, in fact, by a grown-up version of the science-obsessed schoolboy I was a few decades ago, when my own interest in opening up science began. I spent much of my teenage years hanging out ('hiding from the world' might, perhaps, be a fairer description) in the observatory my school was blessed with. A squat brick construction for the most part, it was topped by a glorious metallic rotating dome, under which sat a large and impressive telescope. A frame made of blue aluminium rods supported a mirror fifty centimetres across, an impressive size then and still large by amateur standards today. (It was, for example, larger than the telescope that sits on top of the building in Oxford in which I now work, used for teaching undergraduates how to handle a modern instrument.) The size of the telescope presented challenges. When it was pointed straight up at the zenith, any observer was required to stand, usually on one leg, atop a stepladder and lean across in order to reach the eyepiece. The telescope was also slightly too large for its dome, the result of a sudden glorious rush of blood to the head that had led the staff responsible to buy something much larger than they had originally planned, and so any attempt to look low in the North required some acrobatic leaning out over the stairwell, something that added to the excitement of any observing session. The whole thing was controlled by a rather rickety old computer (a BBC Micro with a very sticky '4' key) and was the pride and joy of the Physics Department's staff, many of whom had spent years fundraising for such a magnificent facility.

The ringleaders were head of physics, Graham Veale, and the physics technician, Ian Walsh, along with their friends from the local Torbay Astronomical Society. They'd raised the money partly by running discos for local teenagers, and it still staggers me even now that those who'd endured such things for fundraising purposes would then turn around and hand the keys to a bunch of 12 year olds, but they did, and along with a couple of friends I set out to do some Proper Science.

Not that we got very far. Cloud was a problem, the fact that we discovered that pizza delivery companies could be persuaded to find the observatory was a distraction, and the task of lining up a faint object on the tiny chip of the digital camera attached to the telescope remained almost entirely beyond us. Nonetheless, I remember very clearly the sense that despite the pathos of our limited efforts we were embarked on something important. Something, in fact, that might add just a little to humanity's understanding of the Universe in all of its glory. We were, it seemed, only a piece of good luck away from making a discovery.

The closest I ever got wasn't, as it turned out, at the observatory, but at home, using a much smaller telescope I'd managed to scrimp and save for. The house I grew up in was away from the main road on a quiet cul-de-sac, making the front drive a reasonable enough place to set up for observing, especially once the streetlights were off. A 14 year old immensely proud of a newly acquired telescope, I was out early one spring evening, taking advantage of the warmer weather to get a last glimpse of my favourite object as it sank into the evening twilight.

The object in question was the Orion Nebula, a vast complex of shining gas and silhouetted dust in turmoil as, deep within it, stars are being born. William Herschel, the discoverer of Uranus and a pretty good writer as well as a sharp observer, described it as 'an unformed fiery mist, the chaotic material of future suns', and even through my modest telescope I could see what he meant. Glowing with a gentle, green light, the three-dimensional structure of the nebula was clear, and I would go on to spend hours waiting for those moments when the sky suddenly stands still, taking advantage of momentarily good seeing to try and tease out fainter and fainter details.

Patience helped, as did allowing my eyes to get used to the dark. I also spent a lot of time practising what astronomers call 'adverted vision'—the technique of looking out of the corner of one's eyes in an attempt to use the rod cells which lie there and are particularly sensitive to faint light, rather than the more central cones which specialize in colour but are less good when the going gets tough—but it was still a challenge. My telescope, pride and joy that it was, was funded via a weekend job selling buckets and spades to tourists bound for the beach, and it couldn't be described as sophisticated. One of its special features was to display, unless it was repeatedly nudged, a sad tendency to slump slowly towards the floor.*

* I have yet to get this fixed. I should get round to it, I suppose.

On this particular evening, I'd become distracted by something on the street and had let the telescope slide away from the nebula. When I returned to the eyepiece, framed perfectly in the field of view was what seemed to me to be an especially beautiful star cluster. There were maybe twenty or so stars, of roughly equal brightness, with hints at the edge of visibility that many more existed. Surprised I'd never heard of such a glorious cluster close to a tourist stop as popular as the Orion Nebula, I looked it up in my star atlas and found, where the cluster should be, nothing but a blank space. In those days before the internet was available at home at the flick of a screen there was no way for me to immediately investigate further, and I proudly picked up a pencil (not a pen—I obviously wasn't quite that confident!) and marked its position very carefully on the atlas, next to a carefully calligraphed label: 'Lintott 1'.

I went to bed that night not really believing that I'd made a discovery, but I couldn't stop a tiny part of my brain thinking that there was at least a chance that I might have done. Of course, when I made it to the internet* the next day the cluster turned out to be well known. Its true name, NGC 1981, doesn't have the sonorous ring of Lintott 1, but has the advantage of having been in use by astronomers since the late nineteenth century (Plate 1). Minor though it was, the incident stuck with me as my closest contact with a long-ago epoch when anyone with time, luck, and modest equipment could make a discovery. The astronomical books and magazines I immersed myself in were full of tales of amateurs discovering enormous storms on Saturn, spectacular comets, or the twinkling of variable stars, but they did so partly because they had kit beyond my wildest dreams. People like

* People younger than me may be surprised by the idea that the internet wasn't immediately available from home. It was a bleak time.

retired telecoms engineer Tom Boles of Suffolk brought enormous dedication to their searches, in Tom's case resulting in the discovery of more than a hundred supernovae (explosions that mark the death of massive stars), but they also depended on tens of thousands of pounds' worth of automated telescopic power. These super-amateurs were making spectacular progress, but they were doing so by effectively professionalizing their astronomy, often spending a fortune on advanced kit. The market for buckets and spades being unlikely to expand to a sufficient extent, I felt that finding even something as quotidian as an undiscovered asteroid—once so commonplace as to have attracted the reputation of celestial vermin—would be forever beyond my reach. My romantic vision of amateur astronomy with a small telescope dead, I moved on to become a professional astronomer.*

Here, as with the amateurs, I found a community and a way of life that was under severe assault from the onrushing forces of technological advance. While once individual universities or particular countries aspired to have their own research-grade telescope, a severe epidemic of 'aperture fever'—the desire for larger and larger telescopes—has hit the astronomers of the world, leaving us with no option but to pool resources in ever larger and more expensive facilities. For most of the twentieth century, the world's largest working optical telescope was the 200-inch Hale Telescope at Mount Palomar in California. Built

* That's not to say that there isn't anything worth doing with a small telescope. I've had hours of pleasure staring at the Moon, Jupiter, the Orion Nebula, and countless other objects. If you've never looked through a telescope, find your nearest astronomical society right now, go along to a public observing night, and get them to show you the sights. It's more than worth it, even if you don't get a comet named after you. There really is something special about having photons from far away objects hit your retina, something which just can't be replicated by looking at even the most impressive photos.

during the 1930s and then commissioned in the immediate aftermath of the Second World War, the 200" is a beautiful relic of a bygone age, to the extent that the original design's reliance on a newfangled technology—welding—was considered risky when it was completed. The welding worked, and the telescope was unrivalled until the 1990s.*

Since then, a host of telescopes with mirrors eight metres (307 inches) or more across have been made, and astronomers are working on three separate projects that take us up to the equivalent light-collecting area of a mirror thirty metres or more across. One of the projects, now known logically enough as the European Extremely Large Telescope (EELT), started life as a concept called OWL—the OverWhelmingly Large Telescope. That was supposed to sport a mirror one hundred metres across, and it's been suggested they could have kept the acronym but have it stand for Originally Was Larger. The EELT will still be a monster. Its dome, for example, is the size of a sports stadium, but competition to use such an enormous facility when it's completed in the mid-2020s will be intense. With fewer large facilities available in each generation to be shared among the growing community of astronomers, the pressure to make the best use of every second of observing time has grown, and that's had serious consequences.

My PhD work involved several trips all the way from University College London in the heart of Bloomsbury to the Big Island of

* A larger telescope was built by the Soviet Union in the mid-1970s, but it was plagued by problems bad enough for most to discount it in the competition for the world title. The original mirror was so bad it had eventually to be replaced, but even then the telescope remained almost neurotically sensitive to changes in temperature. That being said, with a lot of help from those who know it well, my collaborators and I have managed to get good data from it, so perhaps writing it out of the history books is a tad unfair.

**Figure 3** The summit of Mauna Kea in Hawai'i, home to some of the best conditions for astronomical observing anywhere in the world.

Hawai'i, with the aim of taking advantage of the crystal clear skies available to observatories on top of the (hopefully) dormant Mauna Kea volcano (Figure 3). It is a spectacular and otherworldly place, an hour or two's drive from palm-lined beaches yet regularly covered in snow. It is, in fact, the summit of the world's tallest mountain, double the height of Everest when measured from its summit more than 10,000 metres above sea level to its base deep in the ocean, and high enough that the low air pressure at the top presents a serious obstacle to clear thinking—an environment so hostile that the only thing that lives on the summit is a species of beetle found nowhere else on Earth. I've always wanted to see a Wēkiu bug, as they're known—a strange creature which sits and waits for food in the form of smaller insects to be blown up to it by the winds that sometimes sweep across the summit, and is surely evolving to consume astronomers—but they are very rare.

Mauna Kea must be a strange place to spend most of your days. I wasn't too surprised to discover that several of the staff had been submariners in their previous life, and were therefore presumably acclimatized to spending lots of time shut into metal boxes in a dark and hostile environment.

Whatever it's like to spend a significant proportion of your time there, it was a glorious place to visit, and I loved leaving the observatory buildings for a glimpse of a Mars-red dusty landscape high above bright blue sea or, better, for a spectacular night sky illuminated only by the stars and, just occasionally and far below, the flickering that marks the flow of magma from the still-active Kīlauea volcano into the ocean. (Astronomers have had a slightly more exciting time with Kīlauea recently, as the major eruption in the summer of 2018 produced a cone of sparks that could be seen from the summit as well as producing warnings of shut airspace and toxic vog, a volcanic fog, across the island. I know at least one survey that's six months behind because of delay caused by fear of volcano-induced earthquakes.)

The dream of travelling to this amazing place, and being there on a scientific mission with purpose, was a large part of what attracted me to being an astronomer, but things are changing rapidly. To gain access to the telescope that I used in Hawai'i, one bids for access, competing against other astronomers to explain the (scientific) justification for being allowed to take the controls. This can be tricky, requiring one simultaneously to argue that the results will be transformative, and that you know what they will be sufficiently well that you can promise the time won't be wasted.

Still, if successful then the telescope was yours for the night and all you needed was a bit of luck and a clear sky. If conditions were lousy (and too much time spent under beautiful Hawai'ian skies tends to lead to over-optimistic planning or some very picky observers who want nothing but the calmest, stillest nights) then there is nothing for it but to turn round and apply again. Desperation, not to say depression, could set in; I remember observing with a radio telescope while rain swept across the mountain top, washing away any chance of decent data, and on another occasion nearly a week spent kicking pebbles along the gloriously palm-swept beach in high dudgeon at the sheer unfairness of cloud.

Worse from the observatory's point of view are the nights where I got lucky, or at least luckier than I should have. My PhD was mostly on the subject of astrochemistry, and I'd got time to go looking for molecules in space using a dish fifteen metres across known as the James Clark Maxwell Telescope, or JCMT. The JCMT works in the region of the spectrum that astronomers call the 'sub-mm' which and everyone else calls microwaves, which explains why it was high on the summit of a volcano. Your microwave oven works by firing waves at a wavelength chosen to excite the water in your food; the water in the atmosphere emits radiation at similar wavelengths, so from ground level the microwave sky shines brightly. By climbing a mountain we could get above most of the atmosphere's water and see clearly out into the cosmos.

I was searching for chemicals like hydrogen cyanide in and among the clouds of star-forming regions like the Orion Nebula. That isn't quite as quixotic a quest as it sounds—the surfaces of dust grains in star-forming nebulae provide excellent sites for the sort of chemistry that forms complex molecules, and observing them provides more information than physical measurements alone ever could—but because for the most part we were happy with a single detection, rather than needing a map, a blurry view would do and these observations could be completed when conditions weren't great. Nights with the best 'seeing'—those when the air is crisp and still—would be better spent on high-resolution mapping or imaging projects, but if one of those came along when I was on the telescope then it was going to be used for detecting hydrogen cyanide no matter how good the skies became.

When both the astronomer sitting miserably on the beach and the observatory management started wondering how to make their multimillion dollar instruments more productive, it was only a matter of time before we found our way to a more efficient means of doing things. Gradually, over the past twenty years, the major observatories have shifted their operations to a model in which observations are dynamically scheduled; if the conditions improve, then high-priority objects which require the best weather can be targeted and previously scheduled work shelved temporarily. In most cases, it's become rare to travel to a telescope, and rather than carrying back precious images in triumph from Hawai'i, or Chile, or the Canary Islands, the arrival of fresh data is signified by the ping of an email hitting my inbox. I miss my observing trips, but I worry more about the next generation of students who won't necessarily have any hands-on experience with where their data come from.

A healthy respect for data is critical in developing scientific scepticism, but once the astronomer was removed from the process of observing it at least became clear that new ways of working were possible. Whereas astronomers like me are used to observatories, general purpose facilities that are built for a multitude of tasks, what we're getting increasingly in astronomy are experiments; there's a tendency to move away from targeted observations altogether and towards collaborative surveys of large chunks of sky. Such surveys produce ever-larger repositories of data held in trust for hundreds or even thousands of scientists to use, none of whom need have gone anywhere near the telescope upon which their research depends.

The most successful by far of these projects is called the Sloan Digital Sky Survey. It's named for and funded by the foundation set up in the memory of Alfred P. Sloan, the man whose modern and, above all, efficient management techniques turned General Motors into the colossus that bestrode mid-twentieth-century America. He would have approved, I suspect, of the business-like way that his namesake fulfilled its primary mission, scanning the sky from its home in New Mexico in order to create a three-dimensional map of the Universe including nearly a million nearby galaxies (Figure 4).

Is a million galaxies a lot? A single, medium-sized galaxy like our own Milky Way contains roughly one hundred billion stars, and so Sloan's sample contains plenty of interest. Yet there are at

**Figure 4** The Sloan Digital Sky Survey telescope in New Mexico. The main mirror is 2.4 metres across.

least one hundred billion galaxies or so within the span of the observable Universe—as many galaxies, in fact, as there are stars in the Milky Way—and so this, the most detailed map we have of our surroundings, only includes 0.001 per cent of the available sample (Plate 2). Even by the standards of Earth's medieval map-makers, that's a lot of undiscovered country marked 'Here be dragons'.

Yet the effort isn't hopeless, and another analogy might help. Election pollsters in the US are faced with predicting the behaviour of something like a hundred million people, yet typically use a sample of maybe a few thousand voters who they can reach on the phone or online. The ratio between observed and estimated samples is almost the same, and so we might conclude that we should expect our knowledge about the Universe based on the Sloan observations to be about as accurate as a single opinion poll. Actually, things aren't that bad. Not only are galaxies much simpler than people (a fundamental truth that makes astrophysics possible), but the Universe is much less variable on large scales than is opinion in America. For most purposes, therefore, we can assume that the volume covered by Sloan is a typical chunk of the Universe, and draw conclusions about the whole based on what the survey shows us.

One can, of course try and look deeper into the Universe as well. One of the great advantages astronomers have in trying to understand the Universe is our ability to see into the past. Sloan is a survey of the local and thus the present-day Universe, but as we look further away we receive light which has taken billions of years to reach us. By combining our views of near and far we can piece together the whole story.

The main scientific goal for which Sloan was built was to study the carefully plotted positions of its million galaxies in order to measure the expansion of the Universe. The galaxies are, in this kind of study, nothing but trace particles, carried along on the grand expansion of space like so much flotsam carried by a river in flood. The first step towards this grander goal, though, was to identify enough galaxies, and that meant imaging a large area of the sky; in fact, Sloan ended up covering about a quarter of the entire celestial sphere. For most of the time, the telescope followed the simplest possible observing strategy, allowing the sky to turn overhead while its sensitive camera grabbed images of whatever passed across its field of view. In all, more than 300 million separate objects were recorded by the survey in its eight years of operation, and the resulting database and the pile of pictures that accompanies it are uniquely valuable. Among the haul, nearly a million fuzzy objects were identified as galaxies which were likely large enough, bright enough, and above all close enough to allow us to discern their structure. For many of these galaxies, we had not only images but spectra, careful studies of their light at each wavelength, which revealed the distance to the galaxy and much else besides.

This all sounds pretty impressive—and it really was—but the really groundbreaking thing lay in how the team of thousands who dedicated years of their lives to designing and operating the Sloan survey treated the precious data that resulted from their efforts. They would have been perfectly within their rights and consistent with historical precedent to hoard it for their own private use, taking the time to publish paper after paper while safe in the knowledge that, without data to match, the rest of us had no way of scooping them. Yet to my continued astonishment, they chose to share the fruits of their labours with the world; an astronomer like myself who had put in no work at all has exactly the same right to use the data as those who had spent every working—or waking—moment of the last decade or two dreaming of what it might reveal. Not surprisingly, the Sloan data quickly became one of the cornerstones of modern astronomy, triggering and then fuelling an explosion of interest in studying galaxy formation and evolution, a field of study that holds the key to understanding the history of the Universe.

I stepped into the changing world of professional research when I spent a summer as a 17 year old at the University of Hertfordshire, sponsored by the wonderful Nuffield Foundation to take six weeks to experience what life as an academic was really like. Nuffield still sponsor thousands of British students to spend time doing research during the summer, an experience I highly recommend to anyone thinking of research as a career, and looking to do something independent.

I was nominally employed to look at the effect icy dust grains had on light travelling through the environment about newly formed stars, but in reality this meant running computer programs over and over again while eating an obscene number of Danish pastries obtained from the university's library cafe. The work itself was tedious, and I wasn't very good at it, but I did enjoy the company of the astronomers and a glimpse into their world. Having written up the summer's efforts I found myself at a 'science fair' organized by the British Association, and more through a certain gift of the gab than any scientific skill ended up as one of the UK's representatives at the International Science and Engineering Fair, an annual American jamboree held that year in Philadelphia, and my first introduction to the weird cult of school science fairs that prevails across the pond.

The aim of these events is laudable. Through a hierarchical system of school, city, state, and national science fairs every pupil studying science could have a chance to get to grips with science as it is really practised, not just as presented in a textbook. Science fairs are a big deal in the US, as much a part of the high-school experience as rituals such as the prom (similarly foreign to me), and the competition in Philadelphia was fierce. Scholarships worth hundreds of thousands of dollars were available to prize-winners, and it would be difficult to underestimate the competitiveness of two thousand or so teenage overachievers. I knew I was outmatched as I carefully stuck up the A4 pieces of paper that described my project, watching out of the corner of my eye as parent-assisted competitors assembled fully lit display booths and prepared experimental demonstrations. (In my memory at least, the winner that year was someone who had built a plasma chamber in their back garden.)

When the judging started strange things kept happening. Adult after adult looked at my pieces of paper, and then started asking where my hypothesis was, and how I'd gone about testing it. It's not a completely crazy question, and you'll be familiar, perhaps, with the idea of hypothesis testing from school science—you write down the idea you're trying to test and the alternative, boring, 'null hypothesis', and then use data to distinguish between the two. For a simple classroom experiment, you might have a hypothesis like this:

> Talking to plants will significantly improve their growth rate.

And a null hypothesis like this:

> Talking to plants will make no difference (to the plants—effects on humans are not the focus of this experiment).

You could then take two plants, talk regularly to one while keeping the other in splendid isolation, and in measuring the difference between the growth rate between the two gain some evidence in favour of either the hypothesis or its null partner.*

* I'm no botanical expert, but I did spend some time trying to find out what would happen if you actually did this. I'm sorry to have to report that the scientific literature on this vitally important question is somewhat sketchy, but it

It's harder in astronomy than in basic botany to design simple experiments, but in my case the judges were expecting the hypothesis printed proudly on the first sheet of paper to be something like:

> Scattering of light off dust grains is responsible for the high levels of circular polarization observed in star-forming regions.

The null hypothesis would have been something like:

> Scattering of light off dust grains cannot be responsible for the degree of circular polarization observed in star-forming regions.

According to the science fair judges, devoted to ensuring their competitors headed off to university with a decent understanding of the scientific method under their belt, having written down these formal statements all I had to do was design the right experiment to test them, but I couldn't see that it was that simple. To see why I was confused, I need to explain about the specifics of the problem involved.

Unfortunately, this means understanding the concept of circular polarization, which is both slightly obscure and overcomplicated. For starters, think about light. Since the work of James Clerk Maxwell and the other pioneers of nineteenth-century physics, we've known that light can be described as a wave, which travels through space.*

Everyone's familiar with waves, so thinking of light as a wave sounds simple enough. We're used to ocean waves, where a swell moving towards the shore lifts the water as it passes, and sound

seems that while plants do respond to sound, only loud noises have any impact. Plants, it seems, would prefer clubbing to a nice quiet chat in the pub. If you act on this information by taking your yucca out on a Friday night, do let me know how it goes.

* It can also behave like a particle, but that's due to quantum weirdness which need not distract us here.

waves, where sound is transmitted by atoms in the air knocking into each other. (This is the explanation for why in space no one can hear you scream.) Those early pioneers of physics were much occupied with the question of what sort of wave light could be; it seemed obvious that it would need a medium to travel through, but this isn't true. We now know to describe light as a wave that propagates itself, capable of travelling through even the vacuum of space. Think of it as a bundle of related electric and magnetic fields, each of which oscillates as light travels through space.

In this picture, the components of the light—the electric and magnetic fields—have a direction. They can be oscillating up and down, or right to left, or at any angle in-between, and in most circumstances and from most sources we receive light that is a mix of all possible directions of oscillation. There's no particular reason for a source of light to spit out aligned waves.

If the light scatters off a surface, like the ground, this can change. Such scattering can produce light in which some or all of the oscillations are aligned; we say that it has become 'polarized'. This 'polarization' can be useful; by making sunglasses out of a material that only lets through light oscillating in one direction, we can cut out the scattered light. Using such a material lets drivers can see more clearly, undistracted by light scattering off the surface of the road (Figure 5).

Because stars form deep in the middle of clouds of gas and dust, the light from a newly formed star quickly encounters a surrounding cocoon of dust, tiny particles of carbon or silicon about a tenth of the size of an Earthly grain of sand (Plate 3). These particles scatter the light and cause it to become polarized. My summer dabbling in research was concerned with what happens next. If polarized light is scattered again, then, instead of the oscillations all being lined up with each other, in the right circumstances a large fraction of them will tend to rotate in either a

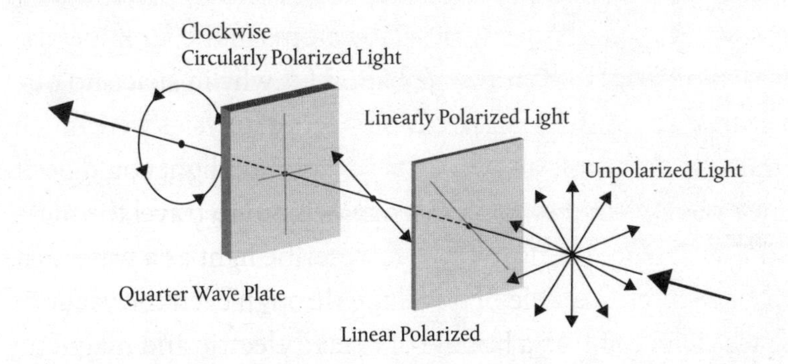


**Figure 5** Schematic showing transformation of light as it becomes polarized. Initially the electric field can appear in any orientation, but after linear polarization there is a preferred direction. Circular polarization favours rotation of the field.

clockwise or anticlockwise direction. This is what's known as circular polarization—we say that light is circularly polarized when we get more clockwise than anticlockwise light from a source, or vice versa.

For most purposes, the presence of circularly polarized light makes little difference to anything, but there is one important exception. Some complex chemicals care deeply about whether the light hitting them is circularly polarized, and as these are precisely the chemicals that life on Earth depends on we too have a vested interest.

This phenomenon happens when an atom such as carbon makes four different chemical bonds, each with a different atom or set of atoms (Figure 6). A bit of thought or a glance at a diagram will show there are two possible configurations, each one a mirror image of the other. No amount of manipulation will turn one into the other, any more than you can rotate your left hand to sit perfectly on top of the same shape as your right hand.

Such pairs are known as 'chiral' molecules, and because they have the same structure—they have the same chemical formula—

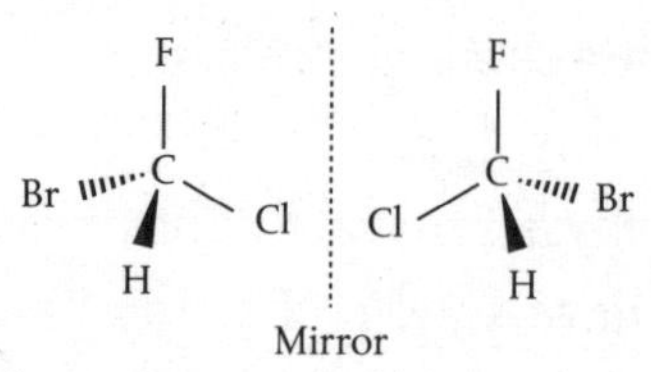


(R)-Bromochlorofluoromethane is not superposable on (S)-bromochlorofluoromethane (its mirror image). These molecules are chiral.

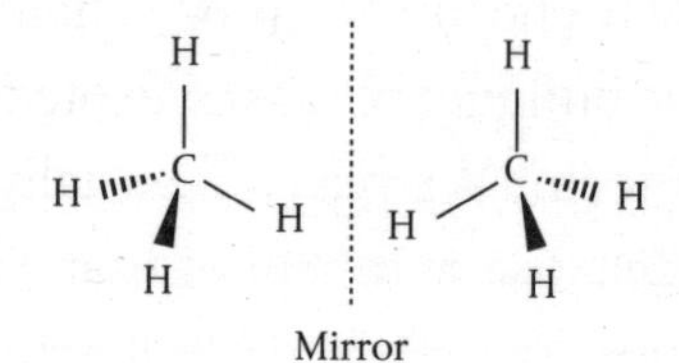


Methane is superposable on its mirror image, and therefore achiral.

**Figure 6** Two forms of bromochlorofluoromethane, which are mirror images of each other. As the central carbon makes four different bonds, one form can't be rotated or transformed into the other.

they will behave the same in chemical reactions unless they encounter another molecule which has this property of having different mirror images. When that happens, then left-handed and right-handed molecules will interact differently. For example, find some spearmint chewing gum; it owes its sickly sweet spearmint smell to the presence of just one type of a mirror molecule called carvone. Swap every molecule of carvone for its opposite and your chewing gum would taste not of mint, but of caraway—caraway seeds have the mirrored form of carvone. That might not seem too bad (I would buy caraway chewing gum, I think) but in some cases a benign compound can be transformed into a deadly poison by the substitution of its mirrored opposite.

Such a case would provide a superb detective plot—it would be difficult for a police chemist with modest equipment to tell the two apart—but what these examples really reveal is that life on Earth has made a choice to prefer one mirrored set of molecules to another. Why this should be is somewhat mysterious, and it seems that astrophysics may have the answer. Recent astrochemical revelations have shown that the chemistry found in and among star-forming regions is surprisingly complex; out in the darkness of space, on the surface of the dust grains from which planets will end up forming, chemicals as complex as amino acids—the building blocks of proteins, and hence of life's chemistry—can form. We haven't actually found amino acids yet, but we've got close and believe that the chemistry is well-enough understood to infer their likely presence.

If such complex chemicals naturally appear in star-forming regions, could these space-forged complex chemicals have been the building blocks for life on Earth? Perhaps. We suspect that the Earth's early years were rather unpleasant, with a temperature on the surface that would cause water and any other volatile chemicals that were initially present to boil away. The water that we drink—all the water, in fact, on Earth—may have been delivered here by an immense bombardment of millions of comets and asteroids later in our planet's history. If that's true, and studies of at least one comet have shown that its water is a good fit for Earth's, then it seems likely that a whole molecular cocktail could have been delivered to the then lifeless surface of what was rapidly becoming the blue planet.

This delivery mechanism may explain life's preference for left-handed molecules. If they formed in space, then they may have been exposed to light that was at least slightly circularly polarized. Light which is circularly polarized so that the electric field rotates clockwise may find it easier to excite left-handed rather than right-handed molecules, whereas an anticlockwise polarization might do the opposite. In such circumstances, if you start with an equal (chemists would say 'racemic') mixture of both left-handed and right-handed molecules, you might get chemistry happening in right-handed molecules than doesn't happen on the left-handed side of things. If that happens, then you can see how the products of such chemical reactions—naturally, the more complex molecules—might tend to be more right than left handed or vice versa.

So if our mix of molecules in space was exposed to light with a sufficient degree of circular polarization then even before it ended up on our planet it may have been processed to produce a bias towards left-handed or right-handed molecules. That's why the work behind my summer project was so interesting—it promised a link between the worlds of star formation and astrochemistry on the one hand, and of astrobiology and the origins of life on the other—and that brings us back to the hypothesis I was supposedly engaged in testing. Could a high degree of circular polarization be produced by scattering light off dust grains?

Science fair etiquette suggests that what's needed is an experiment. Astrochemistry can sometimes be done in the lab, but in this case setting up a star-forming disc of gas and dust and allowing millions of years for it to evolve was a little out of my reach. As astronomers have done for more than half a century, in lieu of lab experiments I had to set up a computer simulation. Using a few simple equations (and someone else's code) it was possible to set up an arrangement where light was assumed to have emerged from a young star of a particular size, mass, age, and brightness, to scatter from a first, dense cocoon of dust, becoming linearly polarized in the process. Then, we want the light in our computer simulation to encounter a second dusty structure—perhaps a disc—and to calculate the degree of circular polarization resulting from this second scattering. It's at this point we have to make choices. We have to decide on the size of the disc, and its shape. The strong and tempestuous winds blowing from the young star have likely cleared all the material away from its immediate surroundings, producing a gap at the centre of the disc, so we need to decide on the size of this region too. The dust grains themselves need a composition—are they carbon, or silicon, or a mix of the two?—and we need to work out whether they have ice or whether they are bare. They need a shape. Are they round? Or needle shaped? If the latter, how elongated are they? That matters because nicely needled grains can become aligned in the presence of a magnetic field, and such alignments may lead to further polarization. How strong is the magnetic field? These and many other questions need answering if we are to make progress, and we're already a long way from the simple plant experiment that could be reduced to a single test.

We can, if we choose, run a large number of simulations. Each time, we can keep almost all of the different parameters fixed, altering only one thing for each run. That might work here, for even complicated astronomical questions reduce to reasonably small sets of equations and variables, but for more complex systems this approach will break down. If you've ever been frustrated by a weather forecast, then one of the reasons is that even with some of the most powerful supercomputers in the world it's simply not possible to build a model of the Earth's atmosphere accurate enough to account for everything that observations tell us must be happening in this very complicated system. In the case of our light-scattering dust disc, we also have to deal with the opposite problem of creating a model so complicated that it can explain pretty much any set of observations.

This phenomenon, known as over-fitting, is a serious worry in cases where our ability to think of variables to fiddle with far outstrips our ability to gather observations to test the worlds created inside our computers. The starkest example in the astronomical world is in the argument, now twenty years old, about how to build a computer model large enough and detailed enough to allow us to study the evolution of large-scale structure in the Universe. Building such a cosmological model from first principles, pinpointing and tracking the position of every atom within a cosmologically significant volume of space, is for all intents and purposes impossible. Yet we don't in most cases have the luxury of treating the galaxies as simple point particles, interacting only via gravity, because to compare the results of a model to the real Universe requires including messy phenomena like star formation (what cosmologists like to call, somewhat dismissively, 'gastrophysics') which depend on the behaviour of individual atoms. A computer model, no matter how beautiful, will fail to match what we can see if it can't predict the formation of the stars whose light we observe and so, instead of building a simulation that would require a computer the size of the Solar System, there is a whole industrial complex of scientists spending their careers building what are called semi-analytic models.

The game here is to guess at a set of simple rules that match, even while they don't explain, the behaviour of the system being studied. Maybe a galaxy converts 10 per cent of its gas to stars every billion years. Maybe it's 5 per cent, or 2 per cent. Maybe it is 10 per cent after all, but the process only occurs only when the galaxy has more than ten billion solar masses of gas on hand. Or maybe when it has more than a billion. Maybe a galaxy can convert a certain percentage of its mass to stars, but after 500 million years activity associated with gas falling into the black hole at the galaxy's centre heats up the gas and prevents star formation. Or maybe that happens after a billion years, not 500 million.

With each additional complication, both the list of rules and the list of things that can be altered to provide a better fit to the observations grow. What starts as a simple set of rules quickly becomes a long list of variables, of parameters that can be tweaked to match the computer model to the real Universe. Need more star formation? Turn the knob on the left. Need your galaxies to stop growing earlier in the Universe's history? Push the red button.

I'm being slightly unfair, and I think most would agree that semi-analytic models do a good job of accounting for the observations of the Universe we have today (there are a few interesting exceptions, as we'll see in Chapter 2), but deciding when to add more complexity to the model is a difficult problem. As you make what started out as simple rules more complex, then you should, almost by definition, always do a better job of matching to any given set of observations without necessarily gaining any new insights. This kind of work, where the skills needed involve deep statistical insight and a good gut feeling for the status of your model, is a long way from the science fair vision of a unifying scientific method with a single hypothesis being tested by a single experiment. The best I can do in writing down a simple hypothesis for a semi-analytic model of galaxy formation is something like 'There exists a model I can make from rules which explains the observations we have of the large-scale structure of the Universe', which is hardly satisfying. It's a long way from what I'd actually chose in studying whether light is sufficiently polarized around a single young star to influence the chemistry.

I think that the process being followed here is so different from science fair procedure that you can think of the computer modelling that's become increasingly important in lots of sciences as a whole new way of doing science.

I imagine a band of stereotypical scientists. One sits, dressed perhaps in a Greek toga or covered in chalk at a blackboard, scribbling equations before writing QED in big letters under some world shattering conclusion. They're a theorist, looking for the mathematic underpinnings of the Universe. A second, wearing a lab coat, is surrounded by bubbling test tubes and complex glassware. Their life is spent weighing things, in adding this to that and occasionally putting the resulting compounds into machines that go 'beep' and which spit out graphs. They're an experimenter, testing the theories the other comes up with.

To this motley crew I think we should add a third character. They sit in a darkened room in front of a desk with four or five computer screens on it. Green numbers scroll upwards on at least one of the screens, and they type in a staccato fashion, causing a complex three-dimensional visualization of something to rotate on yet another screen. They are a computational scientist, and modern science needs them as much as it does the other two. (It also needs them to talk to the others, which is perhaps a much harder problem. But that's another story.)

Understanding this change is key to following some of the most high-profile scientific debates of the moment. Our inability to model each atom of the Earth's atmosphere means that belief in the reality of climate change essentially relies on a prediction from a semi-analytic model of the Earth's atmosphere; every time you hear someone claiming that the science of climate change is falsified by the cooling of part of the Antarctic Ocean, or by an exceptionally cold winter they're enduring, then you're hearing confusion about how these categories of scientific thought are interacting.

This picture isn't yet complete. Computer models, though they produce worlds which can be explored, observed, and experimented upon, are really a way of doing theory that suits our digital age. The equivalent observational mode lies in the freeform exploration of large data sets. Take the Sloan Digital Sky Survey, for example. In some sense it was a traditional experiment, with the goal of plotting accurately the positions of galaxies and thus measuring the expansion of the Universe. Yet if you go to the survey website, for each galaxy caught in its gaze you can download maybe a hundred pieces of information. These include sizes, shapes, colours, and brightnesses, and plenty more can be deduced about each system. Is it a member of a cluster? Has it recently interacted with a neighbour? Is its massive central black hole actively feeding on gas, dust, and stars? We can force these questions into 'traditional' experiments, or we can start not with a hypothesis, but by looking in the data for correlations, discovering for example that the most massive galaxies are reddest or that feeding black holes are bad news for star formation. This mode of discovery could be uniquely powerful. Done right, it holds out the promise of not only providing answers to our questions but of guiding us to the right questions in the first place.

This is the kind of promise that gets magazine articles and even books written, and data-driven discovery was labelled the 'fourth paradigm' of scientific discovery as far back as 2009, in a collection of essays under that title published by Microsoft Research to commemorate the life of pioneering computer scientist Jim Gray. The twin ideas of data exploration and 'big data' have attracted plenty of hype, but they are useful in illustrating quite how science is changing.

Imagine, for example, that you're an astronomer at the turn of the nineteenth and twentieth centuries, interested in stars. Through careful observation, your colleagues have assembled a catalogue of observations of many of the brightest stars in the sky. Despite their diligent work, there's not much to go on. Look carefully at the night sky with the naked eye or with a small pair of binoculars, and it is easy to see that stars have different colours. Try, for example, looking at the two brightest stars in the easily recognized constellation of Orion, Betelgeux and Rigel. While Rigel is blue or white, Betelgeux, an enormous star which would engulf Jupiter were it placed in the centre of our Solar

System, appears orange or even red to the naked eye. As well as colour, we can easily measure the apparent brightness of the stars as well.

The breakthrough came when astronomers realized they could use a variety of methods to measure distances to at least the nearest stars. One simple method relies on an apparent shift—a parallax—in the position of a star relative to a more distant background as the Earth moves around its orbit, just as you can make a finger held in front of your face at arm's length jump from side to side by looking at it first through one eye and then another. What measurements liked these allowed for the first time was the conversion of the apparent brightness of a star—how bright it appears to be—into an intrinsic luminosity which reflects how powerful the stars actually are. So with colour, and luminosity, we have a data set we can explore.

Perhaps there's a relationship between the two. In fact, if you plot luminosity against colour on what's now called the Hertzsprung–Russell diagram after two of the first scientists to do this systematically, you find that many stars lie on a rough line, known as the main sequence. Stars which are bluer tend to be more luminous. Those which are red tend to be less luminous, with the Sun sitting on the main sequence somewhere between the two. Once you realize that the colour of a star reflects its temperature this makes more sense; a blue star like Bellatrix in the belt of Orion, thousands of times more luminous than the Sun, has a surface temperature of about 22,000 degrees Celsius—pretty hot, especially compared to the Sun's 6,000 degrees. On the other hand, some of the coolest stars known, puny brown dwarfs, can have surface temperatures which are mild even compared to room temperature (Plate 4).

That this relationship exists therefore reveals that the source of a star's luminosity must also be responsible for setting the other's temperature, but more importantly the fact that the main sequence exists at all reveals that the stars that lie upon it must share a source of power. In fact, all stars on the main sequence are fusing hydrogen together in their cores to form helium, releasing energy in the process, and those which do not lie on the sequence are either protostars still in the process of getting to the point where they can sustain this sort of stable nuclear fusion, or else those which have graduated to other sources of energy, such as the fusion of helium into other, heavier elements. In this discovery from more than a hundred years ago, there is clear evidence of the fourth paradigm at work, as the exploration of stellar data pointed researchers in the direction of the correct theory for stellar fusion. Of course, the full story of how astronomers came to understand how stars are fuelled is more interesting and complicated than the simple version given above, and worthy of a book in its own right. What is important for my purposes is that the discovery of the main sequence provided powerful support for the idea of a single energy source for stars at very different temperatures and with very different histories.

These days, astronomers studying stars have much more information at their fingertips. Most of the objects captured by the Sloan Digital Sky Survey were not galaxies at all, but stars, and a data set with hundreds of pieces of information about each and every one of them is available to researchers worldwide. This rich resource, and those from more targeted surveys, opens up the prospect of new insights into the processes of stellar evolution, but they also make the challenge of data-driven science apparent. We know, because of the work of Hertzsprung, Russell, and a century of astrophysics, that the 'right' thing to do is to plot temperature (or its proxy, colour) against luminosity. Coming in blind, that's not so obvious; Alex Szalay at Johns Hopkins, a brilliant collaborator and a man responsible for much of the data processing that sits behind the Sloan Digital Sky Survey's power, ran an entire research programme with the sole aim of rediscovering the Hertzsprung–Russell diagram among this data. The catch was that Alex's group wanted to do so with their hands off the wheel, trusting in automated searches to identify the signal among the noise. Trying to discover the cutting-edge science of yesteryear among the modern data deluge sounds like a fool's errand, but it's surprisingly tough, emphasizing that new techniques are critical if we're to make the most of the data that we have.

And what a lot of data it is. Sloan seemed overwhelming to astronomers a few years ago, but what's coming down the pipe is truly scary. I had my first glimpse of this future a few years ago shortly after walking onto the pitch of the University of Arizona's football stadium. College football is, in Arizona as in much of the US, something of a big deal, and the stadium is impressive, immaculately tended and seating more than 50,000 fans of the Wildcats. Its real beauty, though, lies underneath the stands, far beneath the cheering crowds, where planners have taken advantage of the stadium's deep foundations to create a stable environment for the university's world-leading mirror-making laboratory. This is the domain of Roger Angel, a now ageing hippie who combines the sharpest of scientific insight with a craftsman's flair and love of tools. In the 1990s, Roger realized and demonstrated how to make enormous mirrors which were nearly hollow, supported by a honeycomb structure and thus much lighter and easier to manoeuvre than would otherwise be the case. They've become the de facto standard for large monolithic mirror telescopes. The only alternative, utilized by the next generation of extremely large telescopes, is to use a mirror that comprises multiple, usually hexagonal, segments, but where possible the simple charms of a single mirror still hold sway.

When I first visited the mirror lab under the stadium, a typical example of its products was laid out on the floor. Like most of Roger's large mirrors, it was 8.4 metres across, a size dictated not by scientific need or even by cost, but by the maximum size that can be easily transported on American highways. I got to scramble on top of it, and tried hard to imagine it at the heart of an enormous telescope swinging around the sky. Since I was there, that mirror has had its turn on the enormous polishing machine, which ground it slowly to the correct shape with almost unbelievable accuracy by the careful application of a black goo called pitch in a process that, degree of mechanization aside, hasn't changed much since Newton's day. The process took months, but at the end of it the main mirror for the telescope that will drive astronomy's new data deluge was ready.

The telescope in question is known, somewhat clumsily, as the Large Synoptic Survey Telescope (LSST). Large it certainly is; with its giant mirror, it can compete with the largest telescopes in the world today. The key word, though, is synoptic. The plan is for the telescope to complete a general survey, scanning the whole sky available to it on average once every three nights, making a movie of the sky. Among the thirty terabytes of data it will produce every night will be discoveries of asteroids whipping around the Solar System, the signs of stellar death in the form of supernovae, and the flickering of galaxies as material falls irretrievably down to their central black holes. Construction of the telescope is now underway, yet astronomers including myself are still struggling to get our heads around the sheer size of LSST data. Even if, for example, you decide you only care about things that change from night to night, you should expect a conservative estimate of a million alerts a night. Filtering that list of events to find those worthy of our attention is essential for

LSST science, but understanding how to do that well requires a research programme of its own.

The LSST telescope is just a few years away, but we can already see even greater challenges on the horizon. The next big international project in astronomy is a radio telescope, known as the Square Kilometre Array (SKA). Rather than being a single monolithic structure, the SKA will span two continents, scattering sensitive radio receivers throughout the emptiest parts of southern Africa and western Australia. Away from the noisy trappings of civilization (and especially pleased to be free of interference from mobile phones), the SKA will listen to the cosmos with a sensitivity never before achieved. The telescope will be so powerful that there are serious worries that attempts to observe nearby sources with it will be swamped by the presence of millions of previously undetected background galaxies, and serious consideration is being given to the feasibility of finding alien airport radar on nearby planets.*

It's the volume of data that matters, though, and here it's hard to find a proper comparison. I could tell you that the SKA will provide as much information in its first week as exists in the five million million million words that have been uttered during the history of humanity. It is certainly an impressive statistic, with the additional advantage of being true, but I'm not sure it helps one really get a grip on what's going on. Does it help to know that the total data rate flowing between dishes will amount to ten

* Initially thought to be a promising source of signals for SKA-era SETI (the search for extraterrestrial intelligence), the consensus seems to be that the fact that we don't know the rotation rate of the planets involved will stymie any serious search. It seems it may be simply too hard a task for us to expect to pick out an unknown signal without knowing when its host planet will be positioned just right for us on Earth to intercept a signal meant for the incoming space plane from Alpha Centauri. Still, the fact that this is even worth arguing about gives you an idea of quite how sensitive this new telescope will be.

times the current traffic on the internet? I'm not sure, but take my word for it: SKA is a project that will live and die on its ability to handle large data sets, and this vulnerability is not confined to astronomy. Whether you're an oceanographer contemplating data flowing from a new generation of Earth-observation satellites or an ecologist carpeting your study area with motion sensitive cameras, you're going to spend a large part of the next decade thinking about data processing.

Of course, scientists have been here before. It's the largest projects, whether the Human Genome Project or the LHC at CERN, that have had to confront the data deluge first. The world wide web was built as a way of sharing information produced by the latter, but the real action happens deep underground in the experimental cavities within which precision engineering brings the beams of particles, travelling in opposite directions around the 27-kilometre-long tunnel at much more than 99 per cent of the speed of light, together to collide.

At the instant particles collide at these sorts of velocities a tremendous amount of energy is released, creating conditions not seen since the first tiny fraction of a second after the Big Bang.* Most of that energy quickly results in the formation of new particles which fly outwards from the point of collision. Many of the new particles are unstable, and so decay into further particles, creating a complicated cascade of debris. It's this shower of particles, some created in the collision and some the

* This statement of course ignores the possibility of alien particle physicists who may have built colliders greater than our own. This is perhaps unfair; any civilization which has grown to at least our puny technological civilization's level has presumably its share of creatures with the same love of banging nature together to see what it's made of that characterizes the Earthly experimental physicist. Still, if they are out there they don't publish in our journals, which is all that *really* counts.

result of subsequent decay, that crash into the successive layers of detectors wrapped around the collider beam and are picked up by the carefully calibrated instruments. One layer might be designed to deflect and thus measure the properties of particles with positive or negative charge, while a final layer might be a calorimeter designed to absorb the energy of particles that make it that far.

By piecing together what each of these detectors find, the scientists can work out what happened in the short time after the collision. When, on 4 July 2012, researchers from two of the experiments at the LHC, ATLAS and CMS, announced that they had evidence for the elusive Higgs boson—what they had actually seen was a repeating pattern of several different cascades of particles which corresponded to what was expected if the Higgs had (briefly) been created. There is no box of bosons in the CERN visitor centre, but the evidence for its existence had been piling up in collision after collision provided by the LHC's collider team to the eager and waiting physicists.

But how did they find those tell-tale signatures in the data? Most events do not produce Higgs bosons. Indeed, the production of such a particle is enormously rare, but luckily by 2012, over 300 trillion (300,000,000,000,000) collisions had been recorded. That breaks down to a rate of around 600 million collisions a second, or 300 gigabytes a second of data, the equivalent of having the entirety of English Wikipedia read to you seven times each and every second. Were you subject to such a cacophony, I suspect you'd reach for the same solution as CERN's scientists. They filter the data they receive from the collider's experiments, throwing out much of it almost instantly and keeping only those events which match a predefined set of triggers. Anything corresponding to a Higgs event, for example, would be snarfled, saved for future Nobel Prize-winning analysis, but more than 99.999 per cent of the data collected by the LHC is discarded within a second or so of being received.

The LHC, though, has never been just a Higgs-seeking machine. Plenty of other experiments are underway, each with their own set of triggers to snatch information from the flow of live events. One of the most exciting for those of an astronomical bent is the search for dark matter, and this is a little different. Dark matter is, we think, the stuff the Universe is made of. All of these atoms, all of these protons, electrons, and neutrons, all of these neutrinos, muons, and more amount to only so much scum floating on a sea of dark matter. It accounts for about 80 per cent of the matter in the Universe, and the embarrassing fact is that we don't know what it is.

Help is, however, on hand. We have good evidence that whatever dark matter is, it behaves as if it is composed of massive neutral particles. You might think of a sea of particles, each with the mass of the nucleus of a copper atom, but neutrally charged so that it can't interact with light. (Such particles are known as WIMPs; weakly interacting massive particles.) If this explanation is accurate, then it seems possible that dark matter particles will be produced in some fraction of the collisions at the LHC. They would likely shun the embrace of both ATLAS and CMS detectors, fly straight through, and thus show up as a loss of energy in the experiment.

That missing energy would be hugely exciting, for it would mean that an unexpected particle, whether or not it turns out to be responsible for dark matter, was being created within the collider. Knowing how much energy was missing (and thus the energy needed to create such a particle) would allow physicists to focus their search and start to pin down its properties. It's possible, though, that the LHC would already have the data that's needed; if the particles sometimes weakly interact with the detectors then in the morass of previously discarded data should be nuggets of gold.* More likely though, if the LHC is producing (as we all hope it will!) some truly unexpected physics, the years of prior experimental runs will count for nearly nothing. If the triggers weren't set to collect the right type of data—if the evidence for dark matter interactions or whatever else has been thrown out with the junk—then there's nothing for it than to reset the triggers and run the experiment again.

This isn't supposed to be a criticism of the LHC. The truth is that their data rate is so extreme, and our ability to capture, store, and sort data so puny in the face of such an onslaught, that they really have no choice but to throw out much of what is produced. They've also set triggers that might catch likely dark matter candidates, but I like thinking about CERN's struggles to do the right thing because they make clear the complexity of modern science, and the decisions that we have to confront when dealing with large data sets. We are a long way from simple experiments with one variable changed each time and into the realms of big computation—despite the hype, we have become reliant on big data.

Yet not all responses to overwhelmingly large data sets need be so alien. Sometimes, the solution is not to reinvent the process of discovery, but instead to look at what scientists have been doing for years. It's just that with more data, you need more scientists. And that, dear reader, is where you come in. You, and everyone you know.

* I don't mean that literally, although when the LHC is not colliding protons it collides lead nuclei, and this has probably produced a residue of gold, albeit not at an economically viable price. It's still nice for modern physics to have fulfilled the dreams of alchemists through the ages, though.

# 2

# THE CROWD AND THE COSMOS

When did humanity become aware of the Universe? Not of outer space itself, and not only of the stars that speckle our local neighbourhood, but of the whole kit and caboodle, the potentially infinite realm that stretches out for billions of light years in every direction? You can make a case, I think, that it was when we first discovered galaxies, or rather when we found that these often enigmatic objects were in fact immense systems of hundreds of billions of stars.

Look at a galaxy through a small telescope, and you won't see any stars. You won't, in fact, see much of anything, just a misty patch of diffuse light. Only when you realize that that light is generated by a vast number of stars, each too distant to be resolved, does the true distance to these objects become apparent. They are revealed as what used to be known as 'island universes', individual travellers separated by vast oceans of empty space. A few centuries after being displaced from the centre of the Universe by the Copernican revolution, discovering that their galaxy, the Milky Way, was nothing special, dealt the denizens of Earth another blow to their collective ego.

That might be depressing, or the vast scale of the stage on which the Universal drama is played out might inspire you. In either case, these discoveries were only the start of astronomers' attempts to understand the formation and evolution of the galaxies. Our attempts to understand how we ended up with the Universe we see around us, and in particular how the galaxies we see got to be the way they are, are driving some of the most ambitious and exciting projects in astronomy. It's been that way for a while, and back in July 2007 I found myself listening to the latest arguments at a conference in Piccadilly organized by the Royal Astronomical Society.

I'm not sure what images the mention of a scientific conference will conjure up. Maybe a bearded Russian theorist mumbling nearly incomprehensibly at a chalky blackboard? Maybe a whole gamut of scruffy academics engaging in hand-waving debate about obscure and incomprehensible points? The latter's about right, at least for the conferences I go to. Ignore any thoughts about calm and considered discussion; the atmosphere is often more febrile than you might imagine, but the rudest of all are not those indulging in backbiting and snark, but rather those of us who disappear into our laptops, barely conscious of being in a lecture hall at all. We'll have fought for the few seats with power sockets, look up from our iDevices only occasionally, and will—at larger conferences—know to sit on the edges of vast hotel ballrooms so that we can plug in.

To hide from the speaker and get on with work you could be doing at home is rude, and distracting, but I'm as bad, if not worse, than most, and on that sunny July morning you'd have found me prodding at a laptop, grumpy in the middle of the back row in a cramped and already sweltering lecture theatre. It must have taken almost two or three slides from the first speaker to set my mind wandering and the search for viable Wi-Fi to begin.

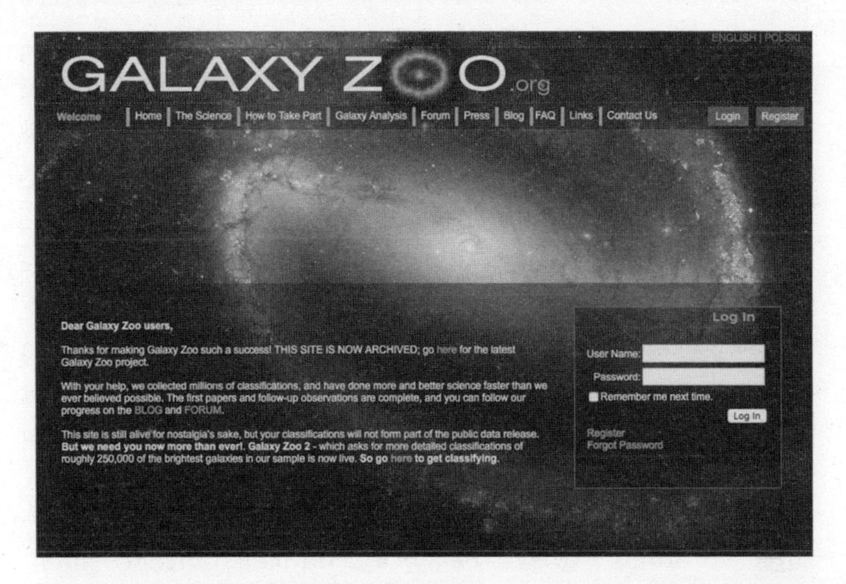


**Figure 7** The original Galaxy Zoo website, complete with Sloan Digital Sky Survey galaxy ready for classification.

At least on that particular day I had a good reason to be distracted. A few hours earlier we'd launched a website called Galaxy Zoo (Figure 7), which asked anyone wandering past to help us sort through the mountain of data that the Sloan Digital Sky Survey had produced.* Having eventually found a connection, I was slightly perturbed that the site wouldn't load. Stranger still, the email address we'd set up (asking people to get in touch if they found anything particularly interesting) was acquiring messages faster than the computer could download them.

The culprit was the BBC News website, where the story announcing our plea for help hovered among the top five most shared articles. We were above a story with the headline 'Garlic "may cut cow flatulence"', which was gratifying for all sorts of reasons, but behind 'Man flies to wedding a year early'. Later in

* The original version of Galaxy Zoo is available at zoo1.galaxyzoo.org and the current version at www.galaxyzoo.org.

the day, we slipped further, as 'Huge dog is reluctant media star' took over at the top, but despite this surge in interest in Samson (at 6'5" Britain's tallest dog but not one for the limelight) the fact that this many people were taking time to check out what was supposed to be an interesting side project was clearly remarkable.

Galaxy Zoo asked people to sort galaxies by shape, a request steeped in nearly a century of astronomical tradition. The galaxies that get all the press, the ones that show up at the start of any science fiction film with a decent budget for special effects, and the ones whose images grace posters, are spiral systems, just like our own Milky Way. We even call these systems 'grand design' spirals, a nod in nomenclature to their spectacular appearance. These celestial Catherine wheels are the Universe's dynamic places, ever-changing discs lit up by the bright blue glow of young massive stars. These stars, the most massive, most luminous, and hottest in existence, burn through their fuel at a much faster rate than relatively puny objects like the Sun. Their youthful presence therefore suggests a galaxy which is still capable of star formation, and they are predominantly found in spiral systems (Figure 8).

A sprinkling of bright stars can mislead. All that's important does not glitter, and to concentrate only on spiral galaxies is to miss the real action. The true heavyweights of the Universe are giant balls of stars which often lurk in the centres of clusters of galaxies. These are the ellipticals. Not much to look at, the reputation of these systems is best summed up as 'old, red, and dead'. In other words, a typical elliptical galaxy is past its prime, devoid of the gas that is the fuel for star formation and missing as a result the young blue stars that give spirals their vim and vigour (Figure 9). These differences show that a galaxy's shape must mean something. Pick at random two galaxies, an elliptical and a spiral, and it's a safe bet that the elliptical will be more massive,

**Figure 8** NGC3338 as seen by the Sloan Digital Sky Survey. This 'grand design' spiral has arms which are filled with clusters of young, blue stars.

have less gas and fewer young stars, and live in a more crowded environment.

In fact, if you are only allowed to know one thing about a galaxy then go for its shape. Its shape—we'd normally say 'morphology' to sound more scientific—contains a history of what's happened to the galaxy over its billions of years of existence, a record of how it has interacted with its surroundings and how it has grown over the years. The division between ellipticals and spirals is really a split between galaxies with different stories to tell, and is as fundamental to astronomers as the realization that humans are broadly split into male and female would be to a researcher studying the health of a human population.

**Figure 9** NGC1129 as seen by the Sloan Digital Sky Survey. This elliptical galaxy lives in a densely populated region, and lacks recent star formation.

This is hardly news. The systematic study of the shapes of galaxies dates back to the most prominent and praised observer of the telescopic age, Edwin Hubble. The man for whom the space telescope is named was originally a scientific outsider, a disappointment to parents who had expected him to go into the family business of law. He got as far as a Rhodes Scholarship to Oxford, but it didn't take him long to realize that the life of a lawyer wasn't for him. After a rather brief stint as a school teacher, he decided at the age of twenty-five that it was time to turn towards his real interest—astronomy.

There's a quotation often attributed to Hubble from this time that has him grandly declaring that he'd rather be a third-rate astronomer than a first-rate lawyer. As it turned out, he did rather better than that. His first astronomical home was Yerkes Observatory in Wisconsin, a strange place to build a facility dependent on clear skies given its climate, but conveniently close to the University of Chicago and at that time one of the world's pre-eminent research facilities. Hubble's PhD dissertation, based on his work at Yerkes and published in 1917, laid the foundations which would support his work for the next twenty years and more. It was a detailed study of what were then called nebulae, a word derived from the Greek for 'cloud' and normally used at the time for a hodgepodge of objects. Star-forming regions like the Orion Nebula I pointed my telescope at as a kid and distant galaxies were both 'nebulae', however different they now seem to us.

For a couple of centuries, observers had added to our catalogues of faint and fuzzy things, but Hubble's contribution went beyond merely finding more of such objects. A telescope isn't a complicated machine, not much more than a bucket for light that obeys the basic rules of optics. One of these rules says that how sharp the images an instrument produces will be (its 'resolution') depends on the size of the mirror or lens being used (and on the wavelength of the light, but that's another story). A larger telescope will, the blurriness and twinkling imposed by the Earth's atmosphere notwithstanding, always produce a sharper image. Despite the Wisconsin weather, Yerkes gave Hubble access to really big telescopes for the first time, and his newly sharpened vision made it clearer than it had ever been that the blanket category of 'nebula' concealed remarkable diversity.

It was a great time to be a young and enthusiastic observer. New facilities were springing up, and after completing a brief stint in the army Hubble found himself in California with access to what was then the largest telescope in the world. This magnificent beast, now known as the 100-inch Hooker telescope, sits atop Mount Wilson looking down on the sprawling city of Los Angeles. The geography of the region conspires to create an inversion layer, with cold air trapped underneath warmer air, which is these days best known for trapping exhaust and producing the smog that blankets that most car-worshipping of cities. If you

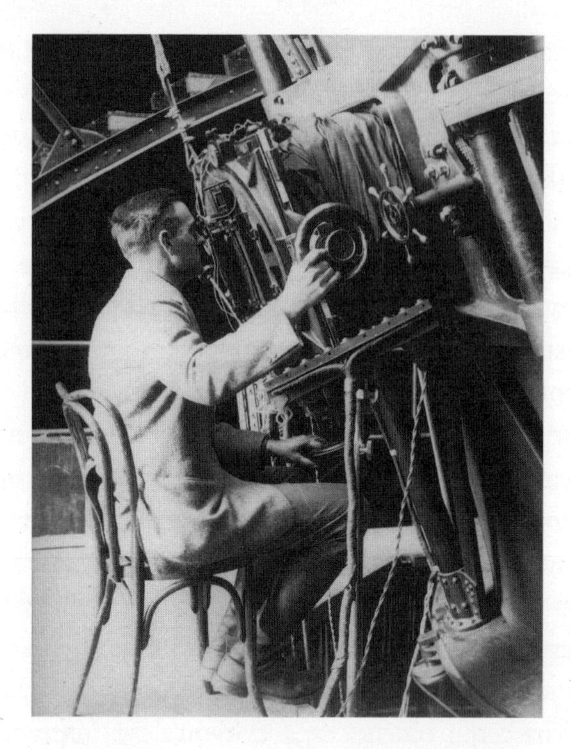

**Figure 10** Edwin Hubble—smartly dressed—observing at the Mount Wilson 100-inch telescope.

can get above the inversion, though, you are rewarded with crystal clear skies, and Mount Wilson is one of the sites with the best seeing in the continental United States (Figure 10).*

It's easy, I think, to imagine Hubble's excitement on arriving in this astronomers' paradise, leaving frigid Wisconsin behind. Whatever his state of mind, he quickly got to work, publishing a

* 'Seeing' is the term astronomers use to talk about the stillness of the air and hence the steadiness of the view provided. There are all sorts of technical ways to measure it, and a few non-technical ones too. For example, at Kitt Peak in Arizona, a count of circling vultures in the late afternoon is a reliable guide to how good a night it is going to be. As a rough guide, the deeper the blue colour of a daytime sky the better the seeing will be; think about the difference between the sky on a hazy summer's day and the deep, crisp blue of a sunny day in winter.

paper emphasizing the distinction between those nebulae which merely reflect the light of stars embedded within them, like Orion, and those which emit their own light. It seems obvious to us now that these latter objects consist of stars, but even telescopes as powerful as those at Mount Wilson refused to reveal individual stars. The obvious explanation is that these systems are far away, but then these nebulae appear bright enough that they must incorporate an almost inconceivable number of stars. This simple chain of logic results in the discovery (or, if you're on the other side of the argument, the extravagant invention) of a Universe of scattered galaxies, each faint smudge of light swimming into view as important in its own right as what has until this moment constituted the entire Universe.

Such grandeur requires equally extravagant standards of evidence, and it's not a bad rule of thumb that any theory which requires a massive reimagining of our place in space is likely wrong. The required clinching evidence that external galaxies really existed soon arrived, thanks to a systematic study of galaxy distances. Measuring the distance to something as far away as a galaxy is not easy, but just as Hubble was beginning his study of the nebulae astronomers had hit upon a useful method which made use of a particular type of variable star—Cepheids.

Cepheid stars swell and then shrink in a regular pattern, and as they pulse they also brighten and fade. That much has been known since the late eighteenth century. They are also relatively luminous, allowing them to be detected in distant galaxies, and catalogues comprising hundreds of the things were quickly assembled. One of the largest was put together by Harvard astronomer Henrietta Swan Leavitt, hired at the college observatory as a 'computer', back when that was a job title and not something that sits on your desk. Leavitt's task was to measure the brightness of stars that appeared on photographic plates obtained by Harvard's telescopes, and she spent particular time on the stellar population of the Magellanic Clouds. These two clouds are satellite galaxies of the Milky Way, in orbit around (and probably being consumed by) our own system, but for Leavitt's purposes they were useful because stars which belonged to the clouds were far enough away that for all practical purposes they could be assumed to all be at the same distance from Earth. So a Magellanic star that appears brighter than another really is more luminous—we don't need to measure its distance, something that causes a lot of headaches when dealing with stars in our own Galaxy. As a result, studying the Magellanic Clouds' stars is key to working out how the Universe is assembled.

Leavitt's catalogues included more than 1,500 Magellanic variable stars, twenty-five of which turned out to show the characteristic Cepheid pattern. They revealed the Cepheids' most important property, an obvious relation between their brightness and how fast they were pulsing. The brighter the star is, the slower the pulse that powers its changes in brightness. This makes some sort of sense, I suppose, as we know that the brightness of a star is partly due to its mass, and it's not hard to imagine ways in which the mass of a star could affect how it would pulse, but it's the use to which this new knowledge could be put that makes it really important. Once the relationship between the brightness and the period of Cepheids is understood, then all you need to do to measure the distance to a galaxy is to find a Cepheid within it. Record the period (the time for the star to complete one cycle of brightening and fading) and you can deduce the distance—a remarkably elegant technique for measuring distances which is as much a part of cosmology today as it was in Hubble's day.

Indeed, a large part of the reason that the *Hubble Space Telescope* is named after Edwin is that one of the high-priority tasks set for it was to find distant Cepheids, expanding the volume of space throughout which we have solid, stellar distance measurements. The experiment it carried out was precisely that upon which Hubble's contemporaries were embarked, and which provided incontrovertible evidence that separate galaxies exist.

And that's not all. Hubble used observations from the new Californian telescopes to show that these galaxies appear to be, almost without exception, moving away from us. The few exceptions that exist are all local. I've already mentioned our Milky Way's cannibalism of the Magellanic Clouds, and the nearest large system, the Andromeda Galaxy, also seems likely to be on a collision course with our own system. In our local neighbourhood, the gravitational pull between nearby systems such as the Milky Way and Andromeda is more important than and can overcome the expansion of the Universe, but on larger scales nothing resists the Universal expansion. What's more, thanks to distances obtained from observations of Cepheids, Hubble showed that the further away a galaxy is, the faster it is receding from us. This observation, now often known as 'Hubble's law',* above all else provides solid evidence of what we would today call the Big Bang. It is Hubble's enduring legacy, although an entertaining debate is underway to decide long after the fact exactly how much of the credit he deserves.

Others had published data sets of similar quality to Hubble, but it does seem to have been his work that captured the imagination, diverting the flow of the debate that was raging in the 1920s and

* As I was editing the book, the International Astronomical Union (IAU) formally recommended that it be known as the Hubble–Lemaître law, to recognize the contributions of Belgian astronomer George Lemaître, who predicted the effect before it was observed by Hubble and others. I am slightly mystified why the IAU decided to busy itself with such a matter, but there was a vote and everything, with 3,167 astronomers in favour and 893 against. You can call it what you like.

1930s over the structure of the Universe. Yet Hubble himself didn't necessarily believe in anything like a modern Big Bang, and, leaving the hard work of building the foundations of the new cosmology to others, turned from using galaxies as particles tracing the behaviour of the space in which they sit to considering them as objects of study in their own right. What he came up with, which can still be found scattered through the pages of today's textbooks, was a tuning fork (Figure 11).

The tuning fork was a way to organize and think about the diversity of galaxies that Hubble observed in the Universe. Along the handle he placed the elliptical galaxies, arranging these otherwise featureless galaxies by their shape. Starting with round galaxies, he worked his way along to those which look like rugby balls, and then to those almost cigar-like in structure. Along the

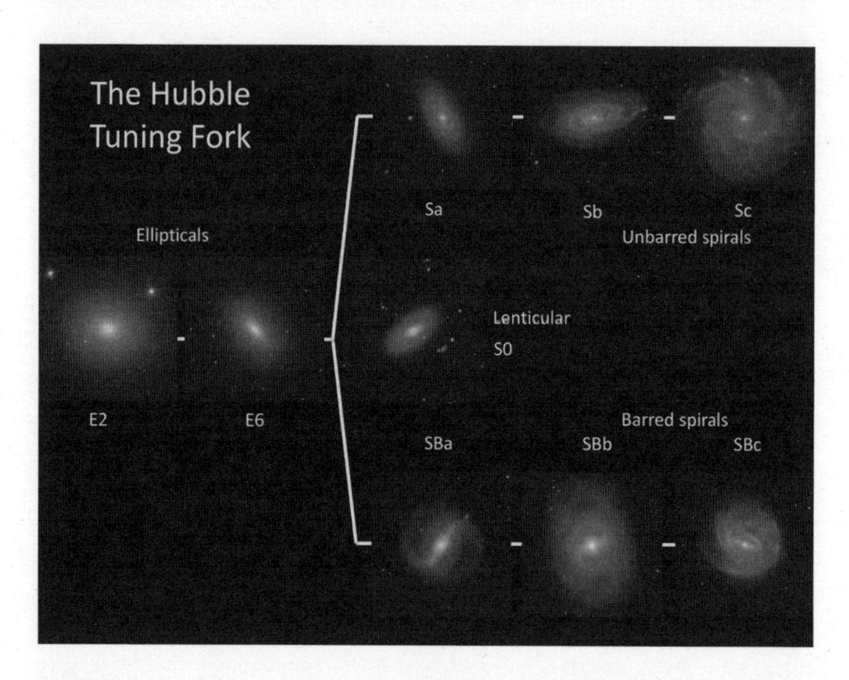


**Figure 11** A modern version of Hubble's tuning fork diagram, still used as the basis of galaxy classification today.

tines of the fork come the spiral galaxies, arranged in order from those with the most tightly wound arms to those where the arms are much looser. One branch was for spirals with a distinct straight 'bar' at their centre—so-called 'barred spirals'—and the other for those without. A few scrappy little irregulars like the smaller of the two Magellanic Clouds apart, such a scheme could account for the whole diversity of the galactic zoo.

What could account for the various shapes? Having sorted them into a nice sequence, it's tempting to see the diagram as an evolutionary one, rather like the familiar lineup that leads from crouching monkey to upright human. There was some support for this at the time Hubble was working, and it seems reasonable to believe that galaxies which started off as featureless balls of stars would slowly collapse under their own gravity to form discs. Within those discs would be spiral arms, the inevitable result of disrupting a rotating system, and these might well then unwind over time, completing the journey from one end of the tuning fork to the other. It's an attractive picture, but one that turned out, sadly, to be complete nonsense. Galaxies don't behave like that. Nonetheless, Hubble's classification still tells us about the shape of galaxies, and it's still used nearly a century later.

The idea that we're talking about classifying galaxies might seem archaic in and of itself. The discoverer of the atomic nucleus and pioneering popularizer of science, Ernest Rutherford, dismissed such work as mere stamp collecting (physics alone, and his kind of physics at that, being exempt from being stamped with his philatelic scorn). But sorting things into categories often marks the first attempt to understand something scientifically, and it can be important even within the inner sanctum of physics, or, in this case, astrophysics.

Those taking a Rutherfordian view might expect classifications based on mere observation to cease to be useful as we start to really understand the physics of galaxies. Simple labelling of what you see is fine to begin with, but Proper Science will proceed differently, and we should expect applying labels as straightforward as a shape to be left as a curiosity in scientific history. Just such an upheaval might be seen to be taking place right now in much of biology, for example, as genomic analysis rearranges our ideas about the relationships between species. A deeper and more useful classification of the tree of life can be found by looking at where the action really is—in the DNA—rather than by carefully observing anatomical features. But for a long time, all biologists had to work with was the power of observation (and plenty would still argue for traditional, rather than genetic work as a means of making progress).

We might imagine that astronomers, too, will find some fundamental measurements that record a galaxy's history and explain its behaviour today. If such fundamental parameters do exist, though, we've yet to find them and thus, lacking any celestial equivalent to genetics, astronomers have little choice, more than eighty years after Hubble, than to carry on dutifully sorting galaxies into categories based on an antiquated tuning fork.

The difference between elliptical and spiral galaxies isn't a temporary one. Left in isolation, a typical spiral will have enough fuel, primarily hydrogen, to go on forming stars for billions of years. As mentioned earlier, recent star formation also means that most spiral galaxies will be blue and elliptical galaxies red; the most massive stars are blue but they are also short lived. This is somewhat counter-intuitive, as you might expect the most massive stars to have more fuel for nuclear fusion on hand and thus to last longer. In fact, the rate at which nuclear fusion proceeds is exquisitely sensitive to temperature, and that in turn depends on the pressure exerted at the core by a star's mass. More massive stars have much hotter cores, and so burn through their fuel much faster, and the brightest and bluest of them may live for only a few hundred million years. Spiral galaxies, and in particular the spiral arms where most star formation takes place, are thus speckled with brilliant blue stars, while ellipticals are, for the most part, red.

Distinguished by their properties, the two sets of galaxies are also separated by their environments. The two types of galaxy often live in separate places. Take the neighbourhood we happen to find ourselves in. The Milky Way and the Andromeda Galaxy are two of the three large galaxies in our Local Group. The Milky Way is a spiral, and Andromeda is also a disc rather than an elliptical. Its exact morphology is somewhat ambiguous, a situation not helped by its nearly edge-on presentation as seen from Earth, with the most prominent feature being a ring of recent star formation. But if we broaden our definitions to distinguish disc galaxies, whether or not they have spiral arms, from ellipticals then the two comfortably sit within the same category.

This reflects the fact that the shapes we see are essentially a result of the dynamics of the material within the galaxy. Disc galaxies like the Milky Way are composed of material which orbits the centre in an organized fashion. The Sun, for example, travelling at a little over half a million miles per hour, completes one orbit around the centre of the Milky Way in just over 225 million years, a length of time that is sometimes referred to as a 'galactic year', and most of its neighbours will do the same. The disc of the galaxy is relatively thin. As Monty Python's 'Galaxy Song' has it, the Milky Way measures 100,000 light years side to side, but is, out by us, only a few thousand light years wide. Think of the galaxy as a large fried egg—ten centimetres across and a few millimetres thick—and you won't be far wrong. By contrast, individual stars in an elliptical galaxy are each travelling on their own orbit around the centre of the galaxy, but they are all inclined at different angles, producing the football or rugby ball shape we see. There is no organized set of orbits as there is in a disc, and this reflects the real physical difference between the two sets of galaxies.*

M33, the third and final large member of our Local Group of galaxies, is also a spiral, and a beautiful one at that. Its multiple spiral arms are in danger of losing their unique identity, so dominated are they by the presence of young star clusters. This sort of arrangement is sometimes known as a 'flocculent' spiral, its appearance somewhat reminiscent of a scattering of tufts of wool. That detail aside, our local neighbourhood's large galaxy population shows signs of segregation: three large galaxies, all of them spirals.

Should we be surprised? The key is in understanding where we live. If you rank the environments in which galaxies are found, from the most densely populated regions to those which are emptiest, you find that we live in the cosmic suburbs. The Milky Way is not a hermit, in splendid isolation, but nor is our patch of space especially crowded. Looking around, this sort of environment is typically dominated by spiral systems, which seem to thrive today in less dense environments. Ellipticals dominate the Universe's cities, mostly hanging out in vast clusters and superclusters of galaxies.

This talk of environments for galaxies would have surprised the astronomers of just a few decades ago. One of the most

* I should probably point out that the use of these terms in the astronomical literature is often confusing, with things being made much worse by the presence of 'S0' galaxies that look like ellipticals but which have a disc hidden in them. People often use the terms 'early-type' for ellipticals and 'late-type' for spirals, the names deriving from the old mistaken idea that ellipticals eventually turn into discs. I'll just use elliptical and spiral here, but really I'm trying to divide disordered systems from those galaxies, typically spiral, where a regular disc is the most prominent feature.

profound and interesting discoveries of the last few years has been that there is structure in the Universe even on the largest scales that we can map. Take the million galaxies of the Sloan Digital Sky Survey, for example. The survey was critically able to make a three-dimensional map of the Universe, recording not only the position of galaxies in the sky but also their distance from Earth. Cepheids aren't visible at these distances, but we can make use of another yardstick—the expansion of the Universe itself.

I already mentioned that Hubble and others provided evidence that the Universe is expanding, and if you work backwards that leads you to the idea that everything began in a hot dense state just after something called the Big Bang which occurred 13.8 billion years ago (give or take a hundred million years or so; the accuracy with which the age of the Universe can be determined still astounds me). By expansion we're not talking about the movement of galaxies through space, but as the expansion of space itself; think not of actors rushing away from each other, but of a stage which itself expands, widening the space between everyone standing on it. That expansion of space, which proceeds today at a rate such that every thirty billion billion kilometres' worth of space grows by seventy kilometres a second, has an effect on the light travelling through it.*

The expansion stretches the light to longer wavelengths, which correspond to redder colours. More distant galaxies captured by Sloan, therefore, look red when compared to their local compatriots, purely because of this redshift. Their spectral pattern—

* In more sensible if less comprehensible units, astronomers would write this as 70 kilometres per second per megaparsec. A megaparsec is 3.26 million light years, or just over thirty billion billion kilometres. In whatever units, this value is known as Hubble's constant, though it isn't constant, but rather something that will change through time as the contents of the Universe act on the expansion.

what you see when you split the light up into its component wavelengths—shifts too. Elements emit or absorb light at particular wavelengths, and we can create a list of those wavelengths by experiment in the laboratory. We still see the same pattern of lines corresponding to elements such as oxygen in the galaxy spectra, but shifted, and so we can compare the measured galaxy spectrum to the standard one and measure the redshift that way.

This measurement contains information about how the Universe has expanded during the millions of years light from a galaxy has been travelling towards us. (When we want to talk about the intrinsic colour of a system, as when considering star formation, we usually have to adjust for this effect.)

Conversely, if you understand the expansion then the redshift becomes a measure of distance to a particular galaxy. The Sloan survey's efforts thus included taking a spectrum for each of nearly a million galaxies, a task made more complicated by the fact that only the best nights, when the air is stillest, could be used for this delicate work. Sloan did, however, have an advantage over previous surveys, in being designed to take hundreds of spectra at once. For each patch of sky it might observe, a metal plate was prepared, drilled with holes corresponding to the position of each galaxy. Into each of these holes, a fibre optic cable was plugged, carrying the light from a distant galaxy on the last few metres of its journey to the instrument that would analyse it.

Compared to the old method of pointing the telescope at each galaxy in turn, this provided for great efficiency, but at the cost of having to complete the laborious task of plugging fibres into the holes in the plate. A rival survey, called 2dF, spent great time and effort producing a fibre-handling robot to do the job. It's a better long-term solution, but Sloan just relied on the efforts of junior astronomers, whose labour proved an easy if unsatisfying solution to the problem night after night and plate after plate.

The results of all of that effort were spectacular. Sloan and similar surveys revealed with clarity what previous data sets had only led astronomers to suspect. The Universe around us is a honeycomb, a cosmic web of clusters and of filaments which wind their way around enormous empty voids. Each of these superstructures is made up of thousands or hundreds of thousands of galaxies; the largest of them, the Sloan Great Wall, is more than a billion light years across.* The fact that there are differences even on these enormous scales has consequences. Jumping millions of light years in any direction from the Milky Way could place you in a very different place, surrounded by different types of galaxies or even, if you end up in a void, with very little company at all.

In the densest parts of these superstructures, a spiral galaxy like our own would look rather out of place. Clusters and superclusters of galaxies are the realm of enormous galaxies which are almost uniformly elliptical. Even smaller examples like the Virgo Cluster, at 54 million light years away our nearest example, have plenty of ellipticals among their population. The densest part of that cluster, centred on a galaxy called M87 which itself weighs in at 200 times the mass of the Milky Way, is almost entirely populated only by ellipticals, with spirals mostly relegated to a few surrounding groups of galaxies which might still be in the process of being absorbed by the main cluster.

Given these facts, galaxy classification may seem to be a simple matter. Ellipticals are red, and live in great galactic cities. Spirals are blue, and live in the sticks. Anyone wanting to select one at

* Studying such a vast structure is rather difficult, and arguments about the Sloan Great Wall still rage. There's some evidence that it isn't one, but rather three different structures superimposed onto each other. The problem is not merely of cartographic interest; understanding the odds of such a large structure forming provides neat constraints on cosmological theories.

the expense of the other could just sort by colour. Pick a handful of red galaxies, and they're likely to be ellipticals. Stick to blue ones, and you'll probably come up with a fistful of spirals. Alternatively, we could take a bunch of elliptical galaxies and expect them to be red, or collect spirals and then look and find they were almost all blue.

This close connection between colour and shape makes it sound like visitors to the Galaxy Zoo website were wasting their time, but things were much more interesting than that. I'd first had an inkling that things might not be simple when, nearly a year to the day before the launch of the site, I'd gone up to Oxford for a job interview. I was coming to the end of my PhD (and more practically, to the end of the grant I'd been awarded to complete it which was paying the bills). I was nervous, and intimidated by the place and the department, and the most critical part of the day consisted of giving a seminar on my work to a busy crowd. I'd rather more of them had been distracted by whatever was on their laptops than listening to me blather on, trying to demonstrate my deep and abiding desire to work on extragalactic astrophysics by reviewing work I'd just completed.

I'd been working on updating a classic, simple model of how the details of the expansion of the Universe affects the galaxies that form within it. Less than a decade earlier, careful measurements had revealed to astronomers that the expansion of the Universe is not slowing down under the influence of gravity, as might have been expected, but rather that it was speeding up. We still don't know why this is happening,* but it has profound consequences for every aspect of our understanding of how the Universe evolves.

* We blame a mysterious force called 'dark energy', which we will meet again in Chapter 6.

The point of my work was to reconsider the formation of ellipticals. This sort of work, more connected to theory than my normal stuff, is fun, but not my natural territory—hence being worried about the talk. As it turned out, I'd barely started talking, and was still feeling exposed, nervous, and uncomfortable, when I was interrupted, loudly and insistently, from the other end of the room.

I'd just been going through the motions of explaining where my data came from and how I'd selected a sample of reliably elliptical galaxies to use. Looking up, I saw that the man interrupting was on the other side of a long table, short, dark haired and smartly dressed. I was wearing a suit (usually a giveaway in academia that a job is on the line), but my questioner was smartly dressed too, the only other person in the room wearing a jacket. I hadn't noticed him when I walked in, and had no idea who he was, but now he was leaning forward, hands steepled together as he tore into the way I'd selected the galaxies used in my study. The problem, it seemed, was that I hadn't been nearly careful enough in picking out a selection of nice elliptical systems. I'd assumed the red galaxies in my sample were elliptical, but, the heckler said, this isn't true. Not all red galaxies are elliptical, he said, and he knew this because he'd looked at them.

I eventually just shrugged and said I'd done what many others had done in picking out galaxies by colour and moved on, but I was distracted for the rest of the talk. I must have said something sensible, though I couldn't tell you what, and a few months later I found myself wandering up the concrete stairs of the Denys Wilkinson Building, home to astrophysics in Oxford, to begin my post-PhD life as a postdoc in Oxford.

My task was to look at how the chemistry associated with star formation might change in different galaxies, and for that I needed to find star formation happening in as wide a variety of galaxies as possible. That meant I couldn't do what I'd done before, and just select red galaxies in the confident belief that most of them would be elliptical. Instead, I found myself chatting to the smartly dressed chap who'd nearly derailed my interview, and who turned out not to be a senior lecturer with a grudge but a precocious graduate student from Switzerland named Kevin Schawinski.

Kevin had demonstrated that it was possible to distinguish ellipticals from spirals, but he hadn't enjoyed the solution very much. Looking at an image of a random galaxy and being able to distinguish fuzzy spiral arms turns out to be something of a human speciality. Just before I turned up in Oxford, blithely throwing my sample of ellipticals together, Kevin had spent a week doing nothing but looking at images of galaxies. He would eventually work his way through 50,000 Sloan images, demonstrating that at this particular task humans still have the measure of computers.

The fact that the software that runs on the lump of matter between our ears can outperform that running on our laptops perhaps shouldn't be too much of a surprise when it comes to a pattern recognition task like this one. The truth is that evolution has left the human race staggeringly well equipped for galaxy classification, albeit as an unintended consequence of making us good at pattern recognition in general. Think about walking through a busy town centre, preoccupied with a shopping list or the cares of the world. You pass people, tens or even hundreds of them, before stopping dead as an old friend or new acquaintance hoves into view. This is the kind of pattern recognition task we all perform thousands of times a day without even thinking about it, but it's still hugely challenging for computers. Yes, facial recognition software has come a long way—my computer makes a half-decent stab at guessing the identities of my few closest friends when they appear in uploaded photos—but only after the investment of hundreds of millions of dollars in research funding. Sitting in Oxford, thinking about classifying millions of galaxies, the best Kevin could do was knuckle down and start looking. The experience left him with a nice clean sample of elliptical galaxies and, as I found out, some strong opinions about the right way to do things.

Kevin was right to be convinced about the virtues of visual inspection. Unfortunately, the research I wanted to do required classifications for the entire survey. In the old days, back when surveys contained only hundreds or thousands of galaxies, this wouldn't have been a big deal, but neither Kevin nor I were keen to spend the best part of five months doing nothing but classifying galaxies. There was a bigger problem, too. Kevin's classifications are all very well, but any truly radical result that came from using the classifications would be vulnerable to the charge that he simply didn't know what he was doing. Don't agree with the classifications? Criticizing the classifier would have been a sensible tactic, and the only way to ameliorate its effect would be to get a second classifier to work through the same set of galaxies, turning months of effort into years.

Clearly this wasn't a sensible use of anyone's time. I tried buying Kevin plenty of beer in the Royal Oak, the traditional watering hole for Oxford astronomers,* but despite this lubrication he wasn't keen on further classification either. Sitting among the grand old beams of the pub, we realized that the only solution available was to call for help. Since I'd been a kid with a telescope, I'd known (and dreamed about) the discoveries that amateur astronomers could make; here was a chance for them to help me.

* For complex sociological reasons, we're now more likely to be found in the Lamb and Flag.

Selfish, perhaps, but the thought did occur to me straight away that this was a way that anyone could contribute, without needing to spend thousands of pounds on a telescope.

The plan, quickly formed, was simple. Leave the pub. Call in a few favours from people we knew who could build websites to get something simple put together. Give talks to local astronomical societies, including increasingly desperate pleas to help with galaxy classification. Say fifty people in each audience, each of whom do 200 classifications each. Give two talks a month for five years, and we should have had everything classified once by 2012. Would it work? Would people really give up their spare time to help with my work?

I was confident this might be a plan crazy enough to work. I'd heard about—and tried to participate in—a project called Stardust@home which had sent tens of thousands of people searching through blurry images of dust grains which had been brought back from a comet by a robotic spacecraft. (I say I tried to participate; Stardust@home had a test you had to pass before being presented with the real data, and I could never do well enough to get in. Though I'd spent several years thinking about the chemistry that happens on them, it turns out that even with the training the website provided I couldn't recognize an interstellar dust grain if one was staring me in the face.) 'If people will look at dust grains', went the logic, 'surely they will look at galaxies'. And look they did. The flood of traffic that prevented me getting to the website on that summer's day back at the Royal Astronomical Society was testament to how powerful the call for help actually was. Volunteers flocked in from all directions; an appearance on the BBC Radio 4 flagship *Today* programme sent us plenty of traffic from London's political classes (if the email addresses I noticed were anything to go by), and an appearance on the Wikipedia home page sent us a collection of people who were used to rolling up their virtual sleeves and getting stuck in online. So overwhelming was the demand that the server which provided the galaxy images, which had been serving astronomers happily for a year or two from its home in Alex Szalay's lab at Johns Hopkins University, buckled under the strain.

That could have been it for the project, but to my immense relief and overwhelming gratitude the team in Baltimore took pity on us and got the server back online. Soon, more than 70,000 classifications were flowing into the Galaxy Zoo database every hour. Better than that, it was clear pretty quickly that the classifications were good, probably close to if not better than Kevin's. But sorting out exactly how good they were would take some effort.*

One dark evening a day or two after the Royal Oak discussion with Kevin, I was sitting at the bar of another historic Oxford pub, the Eagle and Child, with Kate Land, my officemate.† A brilliant cosmologist I'd long since forgiven for beating me to several scholarships, Kate was best known as the discoverer of astronomy's 'axis of evil', an alignment of features in light emitted just 400,000 years after the Big Bang (the cosmic microwave background) that just shouldn't have been there. Before leaving astronomy for a hedge fund, Kate managed to publish a paper

* This book isn't a history of Galaxy Zoo, nor does it focus on web development. But I'd be remiss if the names of Phil Murray, the original designer, Dan Andreescu, the original developer, and Jan Vandenberg, the sysadmin at the Johns Hopkins University who saved the day on our launch didn't appear somewhere. Nor was launch day the only time I had cause to be grateful to Johns Hopkins; a few weeks after launch, we discovered a bug in the code which meant that classifications were being wrongly recorded. Luckily, the problem was interesting enough to be worth Alex's time, and he was able to straighten everything out.

† You might conclude from this part of the tale that pubs are important in British astronomy. You might think that. I couldn't possibly comment.

using newer data that made the axis mostly vanish, but on this particular evening she was exercised by a different problem. Staring into her glass of Zinfandel blush, she told me about a paper which had just appeared on the arXiv preprint server.

While astronomers still publish papers in traditional journals, the main way they're shared is via this thirty-year-old website. In some fields, especially cosmology, papers are published on arXiv for comment even before they are submitted to a journal; it makes for more rapid communication and allows ideas to be bandied about long before they are ready for more formal review. This particular paper, which had crossed Kate's desk because it mentioned her axis of evil, was by Michael Longo. Longo, an emeritus professor from Michigan, was a distinguished particle physicist who had recently become interested in astrophysics, and specifically in the question his paper set out to answer: 'Does the Universe have a handedness?'

To answer this apparently obscure question, he'd looked at a few thousand galaxies in Sloan, selecting the spirals and recording whether they appeared to be rotating clockwise or anticlockwise. (The direction of the arms tells you, in most cases, which way the galaxy is turning; they drag behind the direction of rotation.) He found, surprisingly, that there were more anticlockwise than clockwise spirals, a result crazy enough to scare cosmologists (Plate 5).

If this result is real, it means two things, both of them dramatic body blows against the modern cosmological consensus. First, it suggests that some force is capable of organizing galaxies scattered right across the enormous volume covered by the Sloan survey; Sloan, remember, covers a quarter of the sky. Second, it violates the nearly sacred rule known as the cosmological principle—the idea that any large-scale observation of the Universe should not depend on your position within it—an alien astronomer looking on the same set of galaxies from the other side would see the measurement reversed, so that more clockwise than anticlockwise galaxies appear. Messing with the cosmological principle is bad news; a violation of it means that we can't trust our own view of the Universe.

There was more to it, too. Longo looked for an axis of symmetry in the data, a way of splitting the Universe in two such that (most of) the anticlockwise galaxies were on one side and (most of) the clockwise galaxies were on the other side. In the case of a very strong effect, you might be able to literally divide clockwise from anticlockwise galaxies with a single axis, but even Longo wasn't claiming our Universe was like that. What he had found was that if you took the line that was closest to that ideal case—the line that did the best possible job of dividing those galaxies he classified as clockwise from those he recorded as anticlockwise, then it aligned almost perfectly with Kate's axis of evil.

As the axis of evil that Kate had found was in the cosmic microwave background, it was a feature of the Universe in its very early days. To find that such a feature persisted in the population of galaxies we see around us more than 13 billion years later suggested that we didn't really understand galaxy formation at all. The growth of everything we see would have had to preserve this Universe-scale feature across the aeons, and there's no good explanation for how that could happen. If you believed this paper, then modern cosmology and astrophysics were about to fall apart.

I should probably emphasize that it wasn't despair for the state of modern cosmology that drove Kate and me to the Eagle and Child that night. For one thing, we simply didn't believe the result. Longo simply hadn't classified enough galaxies, it seemed, to be able to make such claims, any more than you could toss a coin twice and conclude on the basis of just those flips that it had heads on both sides. More spiral galaxies, and hence more classifications, were needed. A couple of extra buttons were easy to add to the Galaxy Zoo interface asking volunteers to record the direction of rotation of spiral galaxies, and we could test Longo's challenge to conventional cosmology.

Actually, because it is such a clean measurement, checking for any rotational conspiracy turned out to be an excellent way of testing the Galaxy Zoo classifications. With so many people taking part, we were able to have many people look at each image. In turn, this meant that for each image we didn't just have a classification, but also some idea about how confident we should be. There is a world of difference, as it turns out, between a galaxy which ten out of ten people agree is spiral and one where only six out of ten click the spiral button. After cleaning up the data, removing the classifications of the very small number of people who seemed to have clicked randomly, we were left with 'clean' samples of hundreds of thousands of both spirals and ellipticals.

The former was the perfect set to test Longo's claims, which we were sure were nonsense. You can imagine, therefore, the confusion in the office when it became clear that, despite having a hundred times more spirals available, Galaxy Zoo also had an excess of anticlockwise galaxies. Was this evidence for a universal magnetic field? Was the Universe small and shaped like a doughnut, as one theorist who shall remain anonymous suggested?*

As it turned out, probably not. Before we rushed to publish our Universe-shaking paper, we took the precaution of flipping some of the images in Galaxy Zoo so that people were suddenly classifying mirror images of the galaxies. Had the Universe really

* This topological solution sounds deep and profound, but was, I suspect, a sign of desperation at explaining such a ridiculous result.

preferred anticlockwise galaxies, for whatever reason, then we would have seen a flood of clockwise classifications of these mirrored systems. No such flood occurred. In fact, we still recorded an excess of anticlockwise galaxies. The fault, it seemed, was not in the Universe but in ourselves. Something in the way that the human eye and brain process images makes it easier to see an anticlockwise than a clockwise spiral. It's not that people are seeing a clockwise spiral on Galaxy Zoo and mistaking it for an anticlockwise one, but rather that we just miss the spiral arms in a small proportion of clockwise systems. That slight bias in our perceptions, added up over the classifications received for thousands of galaxies, resulted in something that appeared significant.

Lots of explanations have been proposed over the years for this surprising result. Perhaps it's something to do with the fact that most Galaxy Zoo classifiers read from left to right. We should probably check by asking the same question of Hebrew or Arabic speakers. Perhaps it's something to do with what's known as the silhouette illusion.* An image of a pirouetting dancer created by Flash artist and designer Nobuyuki Kayahara, this figure has the power to distract an entire lecture audience who will be mystified by their inability to agree on the direction of spin. Stare at the figure long enough, and you will see it flip between clockwise and anticlockwise rotation as your brain changes its mind about how to translate this two-dimensional image into three-dimensional space.

Something similar might well be happening as we view the galaxy images, but as astronomers I'm afraid to say that the explanation wasn't nearly as important to us as the fact that we could measure the bias by looking at the differences in classifications between the real images and the mirrored ones, and get on

* <https://en.wikipedia.org/wiki/Spinning_Dancer>.

with science as a result. By doing so, we found that spiral galaxies which are close to each other are most likely to be spinning in the same direction, suggesting that they inherit their spin from the larger-scale structure around them and discrediting an older idea, which suggested that they could spin each other in opposite directions.

Michael Longo wasn't hugely impressed with our results, if reports of people who've talked to him are to be believed. To give him credit, he went back and redid his experiment with randomly rotated images of galaxies, deriving what appears at first glance to be a significant result which has been published in a journal. I haven't had time to dig into the details, but suffice it to say that I'd be very, very surprised if the Universe turns out to prefer anticlockwise to clockwise galaxies. Such quibbles aside, we were pretty happy. We'd managed to detect a subtle effect, and get decent science out of the resulting data. More to the point, it was clear that the volunteers taking part in Galaxy Zoo could contribute high-quality classifications. It was time to go looking for the unusual galaxies that had inspired us to build and launch the project in the first place—including star-forming and therefore blue ellipticals.

The hunt for these systems was now reduced to writing a database query, something that took seconds and which no one outside of a science fiction film can make dramatic. The results, though, were eye-opening; we found plenty of ellipticals with star formation rates that seemed to exceed the Milky Way's. Our galaxy turns one or two solar masses worth of gas and dust into stars each year, but some of these systems were reaching rates of fifty or more. They were less massive than the average elliptical, but otherwise seemed perfectly normal. Best of all, the careful calibration we'd been able to do with the Galaxy Zoo results allowed us to be sure that for most of them it wasn't that we'd

**Figure 12** The IRAM 30-metre telescope, near the Sierra Nevada Ski Station above Granada. It has the best food of any observatory I've been to.

missed spiral arms that were too distant or faint to be seen. These were really star-forming ellipticals, and now we knew they were there we could point telescopes at them.

The telescope of choice was the IRAM radio telescope (Figure 12), the pride and joy of the Institut de Radioastronomie Millimétrique, situated among the slopes of a ski resort in the Sierra Nevada. IRAM is a strange place. In winter you reach the observatory by taking the ski lift among holidaymakers in designer ski gear, but the same climatic conditions that make for good skiing under clear blue skies make it a pretty good site for observing molecules. It's not as high up, and thus not quite as good for sub-millimetre astronomy as Mauna Kea, but the telescope is twice the radius of the JCMT and the food is the best of any observatory I've ever been to. Skiing plans and culinary delights aside, access to IRAM was exciting because it allowed us to measure the amount of molecular gas present in our blue ellipticals. Molecular gas—primarily molecular hydrogen and carbon monoxide, the latter of which could be detected by the telescope—is necessary for star formation, and so this was the equivalent of checking the fuel gauge in the car. If there's plenty of molecular gas, then your galaxy is good to go for billions of years of star formation. If supplies are running low, then it's the end of the road unless new supplies can be taken on board.

Even before grabbing IRAM data, we knew that not all the blue ellipticals were the same. We could tell, by paying close attention to their colours, that activity in most was on the decline, but the time since the peak of recent star formation could be anything from a few tens of millions of years to a billion or so. What we found from IRAM was that the timing mattered a great deal, with a sudden drop in the amount of molecular gas available occurring roughly 200 million years after the peak of star formation.

Something must be happening to switch off star formation in these systems at roughly that timescale, either by heating the molecular gas so it would no longer show up in the observations we were making with this telescope, sensitive only to the emission from cold gas, or by expelling it from the galaxy entirely. More importantly, these galaxies are providing us with a local laboratory to study the formation of ellipticals, a process much more common in the early Universe. Similar, parallel work by other members of the Galaxy Zoo team, particularly Karen Masters in Portsmouth, teased out what red spirals can tell us; these galaxies, too, are undergoing a transition from a star-forming phase to quiescence, though perhaps a less violent one than the galaxies we'd been studying at IRAM.

By measuring the colour of a galaxy, which represents its current state, separately from its shape, which tells us about its longer history, Galaxy Zoo volunteers are providing real insight into the Universe. Because there's a substantial population of each type of galaxy at all colours, we understand their more complex histories. As the indefatigable and indispensable moderator of the Galaxy Zoo forum, Alice Sheppard, pithily wrote: 'Ellipticals are red. Spirals are blue. Or at least so we thought until Galaxy Zoo.' We know things now we didn't just a few years ago, all because hundreds of thousands of people clicked on a website.

# 3

# NO SUCH THING AS A NEW IDEA

My idea of what constitutes recent history might have become slightly skewed after so much time spent thinking about the Universe. I consider the 'present day' to be an epoch lasting at least for hundreds of millions of years, and I spend most of my time worrying about things that happened billions of years ago. I therefore haven't spent much time recently thinking about the end of the Austrian war of succession, one of the many conflicts that rolled across the European continent in the seventeenth and eighteenth centuries. Part of a series of conflicts which were triggered by disputes over the succession to the Habsburg Empire, it was missing from my school history syllabus (as was any knowledge of the details of the Habsburg Empire that shaped the political map of Europe for centuries, or indeed of anything that happened over the channel between the time of the Roman Empire and the First World War). It must have felt more significant at the time. Indeed, so significant was it even for Britain, which one might have thought was nothing but a distant spectator, the end of eight years of war on the continent was enough to spark dramatic celebrations.

**Figure 13** Green Park with 'magnificent structure', ready for the celebratory fireworks in 1749, which Benjamin Robins hoped to use for his study of ballistics.

It is these celebrations, which took place in the summer of 1749 (Figure 13), that account for what we remember of the war today, at least in the UK. The London festivities took place with a soundtrack by the composer of the age, Handel. His *Music for the Royal Fireworks* was a spectacular success then as now, with a dress rehearsal attended by more than 12,000 paying concert-goers, and the premiere in Green Park in the centre of London packed with celebrants. Green Park today is not exactly quiet, existing as it does primarily as a favourite spot for lunching office workers or for tourists who have been exhausted by their task of staring through the railings at Buckingham Palace, but the hubbub on the evening of the display must have been something else.

Among the gathered crowds, one observer in particular would have had especial reason to anticipate the firework display, if not the music itself. His name was Benjamin Robins, and he was a 42-year-old mathematician with a mission, informed by a taste for tackling not obscure theorems but rather the practical problems of the age. His early work, which amounted to mathematical doodling around the foundations of Newton's theory of gravity, had been published already in the *Philosophical Transactions of the Royal Society*, the world's first scientific journal. He would thus have already been known to most of London's scientific glitterati, and by the time of the Green Park fireworks he had been working on military problems for some time.

Back in 1742 he had published a treatise on the topic of *New Principles in Gunnery*, demonstrating for the first time an apparatus, known as the ballistic pendulum, that could be used to measure the velocity of a bullet. The idea was simple, and the execution elegant. By allowing a bullet from a gun to strike a heavy pendulum, and then by observing the subsequent behaviour of the pendulum, the speed of the bullet could be calculated. Almost comically simple as an idea, Robins' design remained in use as a matter of course until the nineteenth century, and it brought its inventor a certain measure of renown.

It must have been a blow, therefore, when he failed in his application to be Professor of Artillery at the Royal Military Academy in Woolwich. A lesser man might have retreated back to the books, away from practical problems, but rather than simply give up Robins renewed his interest in the science of ballistics. Thanks to his pendulum, he knew how fast projectiles could travel, but not the other piece of information needed to predict their trajectories accurately—how high in the air did they get? The firework display of 1749, the apogee of a craze for such shows that had been steadily building over the previous century, provided a heaven-sent experiment for an ambitious student of ballistics. All that one needed to do was observe the same fireworks from as many far-flung places as possible; if the angle at which the rockets seemed to be flying and their apparent height could be recorded at each location, Robins thought he could reconstruct their paths and understand their potential as weapons.

The plan was simple, but Robins estimated that the rockets might be visible for up to forty miles away from the park. The accuracy of his experiment would depend on recording observations from as many places as possible, and even if that problem could be faced then the measurements would not be simple. Close to the launch site the rocket's angle of attack could be estimated by eye, but more distant observers would have to record the position of the rocket as seen against the stars of the night sky or against the height of some distant building. The former method would require astronomical calculations to be completed, the latter would send observers, out after the show was over, across the fields to measure the height of their chosen landmark precisely along with the distance from their vantage point to the chosen reference point.

Covering such a large area with trained observers was beyond the reach of the resources Robins had to hand, and he must have cursed the lack of a willing (or at least biddable) cadre of military cadets that the Woolwich position would have afforded him. Rather than give up he turned, as we did with Galaxy Zoo over 250 years later, to help from a friendly public. Lacking the web, he used the *Gentleman's Magazine* to advertise his scheme, one of a handful of early attempts to do science through what we'd now call crowdsourcing. The *Gentleman's Magazine* was not such a bad choice; it was the first general purpose magazine to reach a large audience. It had already given Samuel Johnson his first regular paying gig, so clearly was home to an audience capable of recognizing intellectual heft and import as well as being widely distributed.

What readers made of Robins' instructions is uncertain. They were clear enough, if a little on the detailed side. Those between fifteen and forty miles of London were instructed as follows (take a deep breath):

> Observing the angle which a rocket, when highest, makes with the horizon, is not difficult. For if it be a star-light night, it is easy to mark the last position of the rocket among the stars: whence, if the time of the night be known, the altitude of the point of the heavens corresponding thereto, may be found on a celestial globe. Or if this method be thought too complex, the same thing may be done by keeping the eye at a fixed place, and then observing on the side of a distant building, some known mark, which the rocket appears to touch when highest; for the altitude of that mark may be examined next day by a quadrant; or, if a level line be carried from the place where the eye was fixed to the point perpendicularly under the mark, a triangle may be formed, whose base and perpendicular will be in the same proportion as the distance of the observer from the fireworks, is to the perpendicular ascent of the rocket.*

As I said, perfectly straightforward, but perhaps lacking in the necessary urgency and oomph to persuade the casual reader of a magazine many miles outside London to go and stand outside to get a poor view of a distant firework display. While I'm adopting the role of a critic and ignoring the historical distance of a couple of centuries, it seems to me unlikely that anyone would go out the next morning with a quadrant they happened to have handy in order to check the height of a building. Uncharitable, perhaps, but given that the pages of the *Gentleman's Magazine* in subsequent months contain no reports from observers within fifty miles of the display, it is surely fair to call Robins' scheme something other than a success.

* *Gentleman's Magazine*, v. 18, p. 488.

There was *one* report received, from a Welshman in Carmarthen who climbed a local hill having heard from the local press that the show from London might be spectacular. I love the idea of this man—identified by his signature only as Thomas 'ap Cymra' (Tom the Welsh)—standing, squinting at the horizon, straining to pick out anything that might be a firework. His patience was, unbelievably, rewarded as he saw two flashes of light, which he estimated to be fifteen degrees above the horizon. Carmarthen is nearly 140 miles from London, so a successful sighting is either unlikely or clear evidence for profligate spending on a show that was intended primarily to entertain metropolitan crowds, not solitary Welshmen. Indeed, Thomas' main point in his report is to complain about the cost of what he thought he saw. To add insult to injury, Robins didn't believe him either, and when he published his results he used only the observations provided by a friend stationed in Clerkenwell, sixty times closer to the action than Thomas, concluding that rockets reached a height of just over 500 metres.

The failure of Robins' call for assistance led him to rely from here on in on carefully controlled experiments, and this early scientific attempt at involving a crowd cannot, it's true, be said to have contributed much to the sum of human knowledge. It was, in fact, about as much of a damp squib as could be imagined, though perhaps that was inevitable given the fate of the display itself. Fog prevented spectators from enjoying a proper view, and the fireworks set fire to the pavilion which had been specially constructed in the centre of the park. Even on a clear day, though, I don't think many would have participated in the experiment; having struggled to find the right questions to ask Galaxy Zoo volunteers, it is nice to think that prospective citizen science practitioners could have learned lessons about writing clear instructions even at this early stage.

Robins' efforts, abortive though they were, are often cited as the first scientific crowdsourcing effort, but there is at least one earlier example. It involves a total eclipse of the Sun that took place in 1715, and no less a figure than Edmund Halley, famous for his eponymous comet and a leading light in much that was good about eighteenth-century science. A total eclipse is a freak of nature, the result of the apparent coincidence that the Moon and the Sun appear the same size in the sky. (I say 'apparent' coincidence because my Oxford colleague Steve Balbus has argued that the tides that result from this arrangement are particularly fortuitous for the development of complex life; if he's right, then rather than advertising Earth's total eclipses as a wonder of the galaxy and expecting aliens to flock to our otherwise modest Solar System to enjoy it, we should expect alien life to grow up only on worlds where total eclipses are common. Rather than an attraction, our total eclipses would be a signpost pointing extraterrestrial astrobiologists in our direction.)

This coincidence means that when, as happens once or twice a year at most, the Moon crosses in front of the Sun as seen from Earth, we experience not only the temporary extinction of the Sun but also get to see its beautiful, tenuous pearly white outer atmosphere, the corona. The view is spectacular—and everyone should try and see at least one total eclipse in their lifetime—but the downside of the close parity between Moon and Sun is that the shadow of the Moon makes only a narrow track on the surface of the Earth. If you're not under the shadow, then all you receive is a partial eclipse—and even a 99 per cent partial is a pale shadow of the experience of totality itself. The narrowness of the track makes a total eclipse one of the rarest of phenomena from the perspective of any particular place; if you stay put, you would be lucky to get one total eclipse in a millennium. The British Isles, therefore, views totality only rarely. The last total eclipse to cross the British mainland happened in 1999, when the track covered Cornwall and South Devon including my then home in Torbay, where it was cloudy, and the next opportunity for British observers won't arrive until 2090.*

Halley was well aware of the rarity of such an eclipse, and he thus viewed the British eclipse of 1715—the first for 500 years—as something of an opportunity. Because the geometry of the eclipse track depends precisely on the positions of the Moon and the Sun, recording the track's dimensions to great accuracy would, he reasoned, allow him to make accurate measurements of the scale of the Solar System. Making such a calculation requires records of the beginning and the end of the various phases of the eclipse, and so Halley prepared to make his scientific observations from the slightly incongruous vantage point of a side road just off Fleet Street in the heart of London, the then headquarters of the Royal Society. He was a good enough researcher to realize that a single set of observations would be sadly lacking, and so in addition to his own preparations he wrote to the professors of astronomy at Oxford and Cambridge to enlist them in his endeavour.

It was somewhat fortunate that Oxford, Cambridge, and London, major concentrations of the scientifically aware, were all predicted to fall within the track of totality. Yet Halley realized that the track crossed a large swathe of the country from east to west, and that restricting himself to just three sites would be amazingly short-sighted. He took action, publishing a map of the track alongside tables of likely timings, distributing it far and wide alongside an appeal for help.

* It's 23 September. I've already marked my calendar, and by coincidence it's a trip to the West Country again as the track covers Cornwall and Devon. The weather prospects are uncertain.

The map was published as a broadside, a sort of single-sheet newspaper common at the time, and was a commercial proposition. Halley worked with a publisher, John Senex, who specialized in mapmaking, and they sold for six pence each (Figure 14).*

Trusting in the ability and efforts of a public he referred to in the instructions he wrote as 'the curious', he asked that anyone who was capable of making suitable observations should send him timings of the eclipse. The leap of faith required for this establishment figure to ask for help perhaps shouldn't surprise us too much; Halley's great genius was in handling data, which he manipulated in order to bring mathematical rigour to his view of the cosmos. Rather than wanting to rely on a small number of 'professional' observations, no matter how reliable, it seems obvious to me that a great datasmith such as Halley would want to have on hand all the available information. After all, he could always decide later how much weight to place on each record he received.

An eclipse is a one-off. Whatever is planned, there is only one chance and a few short minutes in which to execute even the best laid of plans. Given how things went on the day, it was probably a good thing that Halley cast his net widely. The Royal Society party in central London were successful in their attempts to view and to record the eclipse, as were the observers organized by John Flamsteed, the first Astronomer Royal, a few miles down the river at Greenwich, but the university observatories do not come out well from this story. In Oxford, it was cloudy. In Cambridge, poor Reverend Cotes, then in charge of the observatory, was blessed with clear skies. He did, though, have 'the misfortune to be oppressed by too much company' and was thus

* Not a model we've yet considered for more modern crowdsourcing projects, but maybe we should.

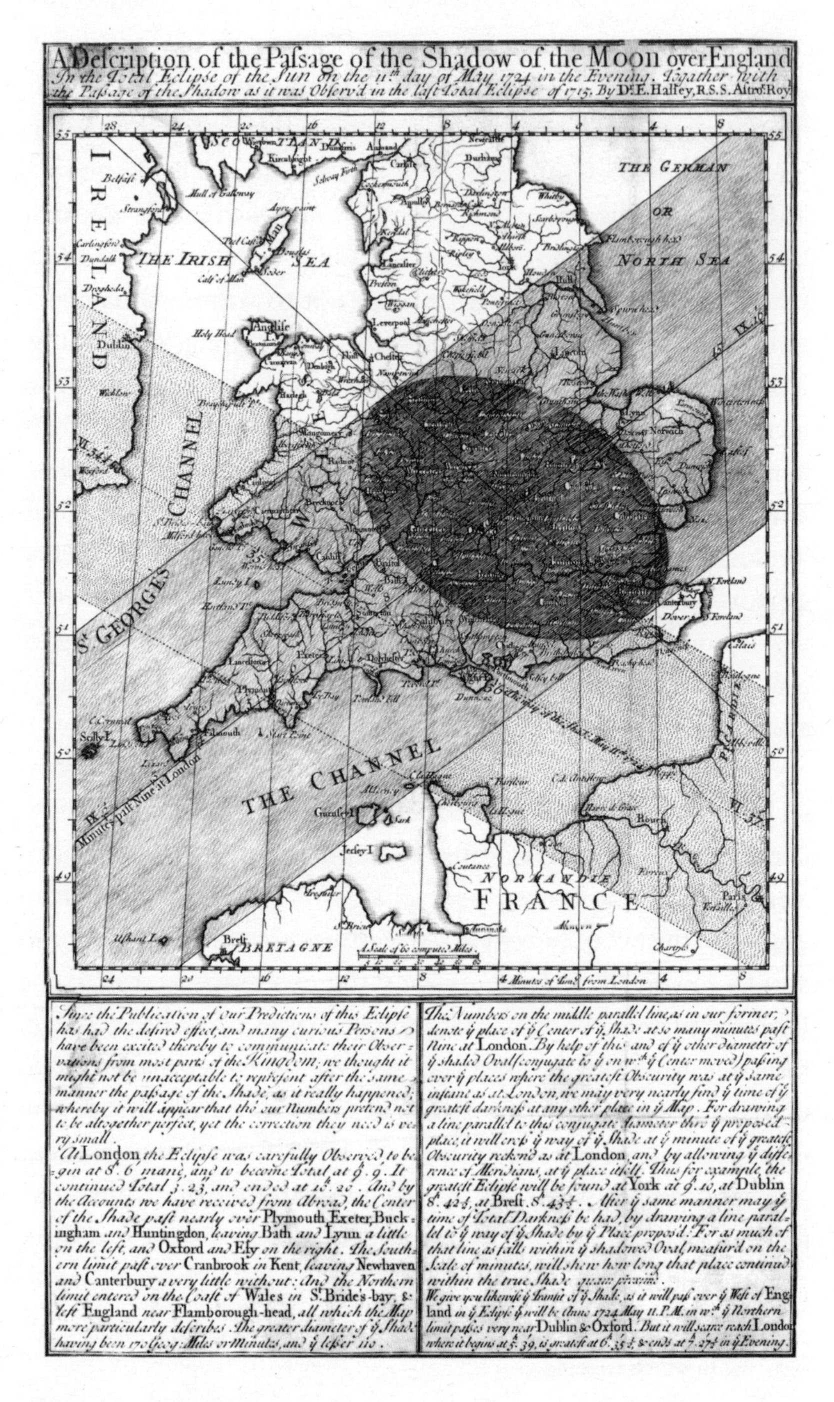


**Figure 14** Halley's 1715 eclipse map, with instructions for making scientific measurements of timings.

unable to contribute anything of use. This seems remarkably lily-livered to me; even if we're charitable and assume that the company in question consisted of crowds of eager eclipse-seekers rather than merely of a friend or two calling round for tea, this distraction seems rather neglectful of the potential for scientific progress.

These varsity failures turned out not to matter, though, for observations flooded in. More than 200 people, as far from London as Plymouth, sent data to Halley. An especial mention is necessary for the particularly heroic souls who eschewed the temptation to head for the centre of the track and a longer eclipse, but who stayed near the edge of the track bearing witness to either no totality at all or a total phase which lasted only a few seconds. A trip of a few miles would have afforded them a more spectacular eclipse, but it was these liminal places that provided the information critical for Halley's experiment. Written up in the pages of the Royal Society journal, the collated results make for an impressive sight, and certainly the 1715 event represented a leap forward in eclipse observation, but ultimately the experiment didn't amount to much. The main result was to confirm that eighteenth-century astronomers already had a pretty good grasp on where things in the Solar System were, and there is no record that I can find of anyone bothering to repeat the experiment when, just a few years later in 1724, a total solar eclipse again darkened parts of southern England. Negative results are, of course, of great importance in the progress of science, marking as they do the roads not taken and the theories not overturned in the search for scientific progress, but it is hard not to be disappointed that this wasn't the start of something more. Rather than standing as spectacular successes which encouraged everyone to reach out to crowds of volunteers, these eighteenth-century stories of Robins and Halley are interesting but lead us to no glorious triumph.

A few decades later, what we would now call citizen science was to trigger an explosion in scientific knowledge, as the age of the grand amateur provided a kick-start to observational astronomy, geology, meteorology, and more. As each of these subjects became established as data-rich and observational sciences, so they came to rely on distributed networks of observers to provide their raw material. That's most clear in cases like that of systematic weather observation, where having data gathered from as many different places as possible makes a world of difference.

The very idea of weather forecasting was somewhat controversial during the nineteenth century, a tale told brilliantly by Katharine Anderson in her book *Predicting the Weather: Victorians and the Science of Meteorology* (seriously—it's one of the most readable scholarly books I've come across in years). In one chapter, she talks about the meteorological efforts of two of my heroes, both of whom built up networks of thousands of weather recorders across the British Isles.

The first, James Glaisher, was an astronomer at the Royal Observatory at Greenwich, back when the site was still supporting cutting-edge research rather than the museum and tourist attraction it is today. Glaisher was a flamboyant polymath with equal tastes for both adventure and publicity, a combination which led to him pioneering the art of scientific study from hot-air balloons in the 1850s and 1860s. Flight then was still a novelty—the stuff of fairgrounds and balls, and passengers clutching champagne—and the distinguished Glaisher who was already, in his fifties, a scientific person of quite some standing who might at first have required some persuading to fly himself. We're told he only took to the air when he became dissatisfied with the work being done by the students and technicians he had deputized to stand in his place. Despite this reluctance, once airborne Glaisher quickly realized the twin advantages of ballooning; not only did it provide access for the experimenter to the atmosphere far above the ground, but it also provided plenty of material sure to attract a crowd to his public lectures. In the process, his transformation from scholar to dashing adventurer was complete, and adventure he certainly had. His descriptions of scientific flights to trace the temperature profile of the upper atmosphere include accounts of episodes in which both he and his pilot all but lost consciousness, becoming 'insensible' and struggling to read their instruments (Figure 15). Even at lower altitudes, the chance of drifting irretrievably out to sea caused constant worry and conflict between the scientist, always persuading himself to take just one more measurement, and the pilot who would have been understandably concentrating on getting back on the ground in one piece.

The balloon used for these ascents was owned by Glaisher's regular pilot, a man named Henry Coxwell, and it came to a surprising end. In 1864, it was ripped apart by a rioting crowd that had gathered in Leicester for a demonstration flight that never happened. Somewhat ironically, it was the sheer size of the crowd that prevented a safe ascent being attempted, but the incident put an end to airborne experiments. The overlap between public display and scientific experiment goes back to the beginnings of Glaisher's scientific career. He had, since 1844, been producing weather statistics as part of an ongoing study into possible connections between disease and the weather. As well as publicizing the results of his investigations in the press, he actively collaborated with the newspapers in order to gather in observations from further afield. London's *Daily News*, for example, helped arrange for station masters to make daily weather reports that could be carried back to the capital and printed in the paper for general edification as well as for scientific use. Others contributed observations too, and contemporary reports are careful

**Figure 15** Coxwell, the pilot, climbs into the balloon rigging as Glaisher lies 'insensible'. Nineteenth-century meteorology looks dangerous.

to explain that the ranks of Glaisher's observers were swelled not only with 'zealous meteorologists' but also with 'the servants and gardeners of landed gentlemen and noblemen'.

The snobbery of that division is striking, at least when we look down from the comfortable heights of a twenty-first-century perspective. All may of course contribute, but only those who do so backed by real means can be meteorologists; others contribute data which can be used in the common cause. This division could be the result of blinkered stereotyping, or it might reflect the simple practicalities of nineteenth-century scientific life, an age the astronomical historian Allan Chapman has called the 'age of the grand amateur', a time when those with the resources to achieve leisure invested it in serious pursuits. There are also signs of a hierarchical view of science: servants and gardeners can make observations but that work isn't really meteorology, which involves analysing the results. Whichever it is, it's amazing to me how clearly these thoughts reflect an argument about what it means to 'do science' that persists right up to the present day.

Once Galaxy Zoo had become an overnight success, we found ourselves casting about for words with which to describe it. Plenty of options were available: crowdsourcing, which I used for Robins' efforts above, was coined by an editor at *Wired* magazine, Jeff Howe, as a portmanteau derived from 'outsourcing to the crowd'. Howe's definition draws on an analogy with 'outsourcing', which is what a company does when it replaces employees who would otherwise be engaged in a particular task with a call to outsiders to complete the task for them. Crowdsourcing would then be what happens when a company or organization asks the world to complete a task for them which would otherwise be handled by employees; a good example might be the recent call by Transport for London, which runs the city's deep tube lines, for ideas as to how to install air conditioning in cramped tunnels.

As with the meteorology example, when you use 'crowdsourcing' it seem that you're making a distinction between those who direct the operation and those who provide the work. Even in the early days, it was clear that Galaxy Zoo seemed to be different; we didn't want to replace Kevin's work with that of the public, but rather to extend the scope of the investigation beyond what could be done by professional astronomers. In any case 'crowdsourcing' seemed a little unambitious, and I found myself leaning towards a more aspirational term—'citizen science'—as something that was more inclusive.*

Rather than placing myself on a pedestal, viewing my position as Eminent Scientist as apart from the crowd—someone who assigns tasks and reserves only for themselves the right to analyse the results—it instinctively seems to me that it is impossible to draw a clear line between where the supposedly menial tasks of data gathering, classification, and exploration stop and some sort of Proper Science starts. It's all just science. Glaisher clearly took much of the credit for the joint enterprise which involved information gathered by observers waiting down the track at remote railway stations, but the value of the whole enterprise is none the less a collective one.

This argument about status and contribution is also visible in the story of the other character that stands out in Anderson's tale of Victorian weather observers. While Glaisher used his status and, somewhat to the distress of his bosses, the name of the Royal Observatory to promote his work, George James Symons

* This term has its problems. The prominence of immigration in political debate, particularly in the US, is a reminder that not everyone is a 'citizen', so 'citizen science' can sound like we're trying to limit who can participate. Whole conferences have been held trying to agree on alternatives, but most suggestions have their own problems. ('Public Participation in Scientific Research', or PPSR, is OK but a little unwieldy.) I'll stick to citizen science, but it is meant inclusively.

had to rely solely on energy and enthusiasm. Symons was the assistant at the then-new Meteorological Office in Whitehall who had responsibility for rainfall (or at least for the measurement and recording of it), and he hit on the same solution to the problem of distributed observation as Glaisher. By 1863, Symons was confident enough to write to *The Times*, inviting those 'of both sexes, all ages and all classes' to send in their observations.

The result was overwhelming, the equivalent of the servers crashing under the weight of Galaxy Zoo traffic. Symons received so many observations that a huge amount of work was dedicated to analysing them, and he quickly became a victim of this success. His boss, Robert Fitzroy (of shipping forecast fame), became convinced that this enormous effort dedicated to collecting and analysing data could only detract from Symons' official duties, and Symons was quickly out on his ear. Short-sighted this may have been, but there is no doubt that dealing with his ever-expanding network must have been terribly time-consuming. Anderson reports that Symons was receiving observations from in excess of 1,000 observers in 1867, and more than 3,000 by 1900. The longevity of the project as well as its scale underscores the fact that Symons had clearly come to regard this as his life's work.

In its early years, the network was supported by a grant from the British Association for the Advancement of Science,* but this ended in 1875. Perhaps realizing the error of its director's earlier decision, the Met Office offered to take on the task of organizing things from there, at which point Symons with some justification told them where to stick it. He felt that the volunteer spirit (and his writings on the subject are everything you would expect

* An organization that is still extremely active, though it is now known as the British Science Association. It was founded in 1831 to promote and encourage science. Its public meetings included the first use of the terms 'science' and 'dinosaur', though not, sadly, at the same time.

from a patriotic Englishman with a cause, writing at the apex of belief in the empire) would be lost or ill-treated if directed not by that same spirit of voluntary contributions, but rather put in the service of some 'Government Office'.

Indeed, he had already been impressively willing to open up discussion about matters of policy to the participants, rather than resting on his own authority. If you want to measure rainfall across Britain, and you rely on volunteers, what should you ask them to do? Their own lives and preoccupations mean that you cannot possibly insist on hourly readings from all. Daily readings seem more sensible, but then when should one make them? Midnight is nice and clean—you'd get a measure of rainfall during a calendar day—but it is hardly respectful of the social lives and sleep of would-be scientists. Symons polled his members, and they decided to observe uniformly at nine in the morning, a nice example of collective experiment design in action. Perhaps it was this collaborative spirit that allowed Symons, following the loss of his grant, to turn to the network's volunteers for funding. Their donations and subscriptions flowed in enough quantity to allow the network to operate at a modest profit during its later years.

Symons and Glaisher provide two early examples of effective citizen science. One created an organization driven by its members who shared a common goal, and who presumably felt part of that greater collective effort. Another used prestige—the Royal Observatory, the *Daily News*—to stimulate an audience to participate by handing over their data. Both were hugely effective, but there is a third, alternative route. In passing, Anderson mentions the story of the Scottish Meteorological Office who, confronting the highlands and islands, felt more than most the need for observers in obscure corners of the country. (My own experience is that such places in Scotland tend to be the wettest, but even then I suppose systematic observation is required to lift such findings much above the level of anecdote.) In 1872, an observer at Stornoway in the Outer Hebrides was recruited by the official, London-based network as a paid contributor. The net result of having the value of his observations recognized by funding was such that he immediately stopped contributing to the existing volunteer-only Scottish network. Pay, it seems, may work as an incentive but a hybrid model was very difficult to sustain. Perhaps the lesson is just that anyone running a project involving volunteers should be very careful not to exhaust the goodwill of those participating.

Due to an unaccountable, or at least unavoidable, lack of contemporary social scientists interested in such questions, we have only speculation as to the motivations of the participants in these proto-citizen science projects. Nor would they have thought of themselves as citizen scientists; the first recorded use of the term in the modern sense is usually given as appearing in the *New Scientist* magazine in October 1979, where 'the citizen scientist, the amateur investigator who in the past contributed substantially to the development of science through part-time dabbling' is mentioned in the context of an article about UFOs. There is an ambiguous reference, dear to my heart, in *Collier's* magazine in 1949, which speaks of 'citizen-scientists' perfecting 'a technique which brought gin to its peak of flavor and high-octane potency', but that seems to be a different thing entirely. Yet though we can't be sure why they participated, one suspects that what united participants was some combination of wanting to belong to a movement, of wanting to advance scientific knowledge, and of rubbing shoulders with (scientific) celebrity.

We get a better sense of what participants themselves were thinking from another great scientific endeavour of the age—the huge and collective burden of keeping Charles Darwin informed and entertained. Observations of the natural world had been part of what we'd call scientific activity for many years before the Victorian naturalist came along, but he was able to draw on a vast network to gather information from around the world, informing his own work while he, after his youthful voyages on the *Beagle*, remained at home.

As that description suggests, Darwin's correspondence, much of which is now available for our enjoyment online, was prodigious. Not quite housebound, but certainly firmly attached to his patch of Kentish soil after his adventures as a young man, he relied on a network of correspondents from every corner of the world to inform him of—well, everything. When I first explored the collection, the first example letter I picked out more or less at random was a note to a Mr Mantell, in New Zealand (Figure 16). In the space of a few short paragraphs, Darwin enquires about some observations of possibly erratic rocks, about whether the Maori ideal of beauty matches that of Europeans, and about the possibility of a creature 'with hair' that was something like an otter or a beaver.

Not all the letters are quite that eclectic, but flicking digitally through the surviving piles gives you a sense of an urgent and vital exchange of information. On the day I'm writing this, 144 years ago, Darwin received a letter from a Mr George Cupples of Fife. George was writing to send his eminent correspondent the 'best wishes of the season', but judging by the rest of the letter clearly felt the inequality between their positions, which he strove to fill with information he thought that Darwin might find useful. The gap is closed by a note on the breeding of Pyrenean mountain dogs, one of which Cupples has recently acquired, and which has 'six well-developed toes on its hind foot'. As if that wasn't enough, a postscript mentions notes on the subject of

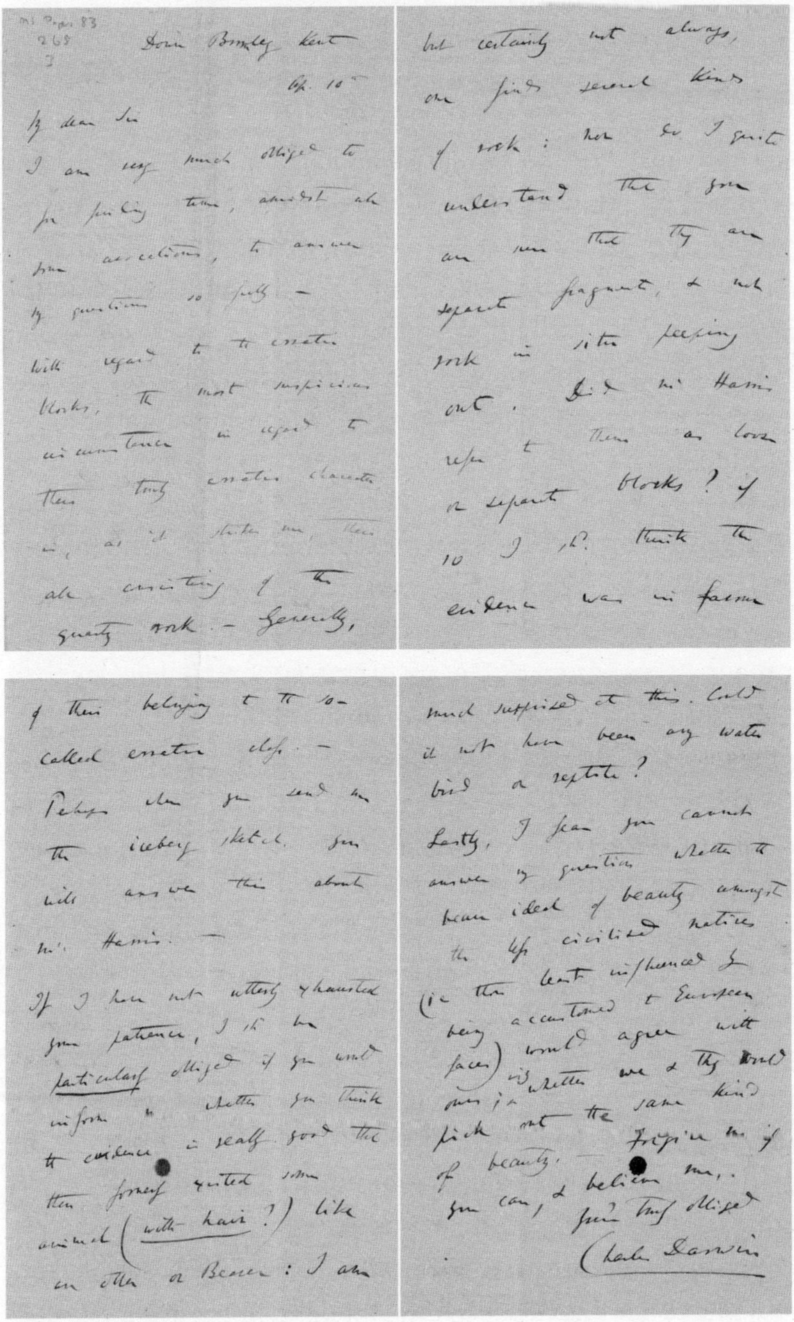

Down Bromley Kent

Ap. 10th

My dear Sir

I am very much obliged to you for finding time, amidst all your avocations, to answer my questions so fully.— With regard to the erratic blocks, the most suspicious circumstance in regard to their truly erratic character is, as it strikes me, their all consisting of the quartz rock.— Generally, but certainly not always, one finds several kinds of rock: nor do I quite understand that you are sure that they are separate fragments, & not rock in situ peeping out. Did Mr Hamilton refer to them as loose or separate blocks? if so I shd. think the evidence was in favour of their belonging to the so-called erratic class.— Perhaps when you send me the iceberg sketch, you will answer this about Mr Hamilton.—

If I have not utterly exhausted your patience, I shd be particularly obliged if you would inform me whether you think the evidence is really good that there formerly existed some animal (with hair?) like an Otter or Beaver: I am much surprised at this. Could it not have been any water bird or reptile?

Lastly, I fear you cannot answer my question whether the beau ideal of beauty amongst the less civilised natives (ie the least influenced by being accustomed to European faces) would agree with ours; ie whether we & they would pick out the same kind of beauty.— Forgive me if you can, & believe me,

yours truly obliged

Charles Darwin

**Figure 16** Letter from Darwin to W. B. D. Mantell, dated 10 April 1856. Sadly, no reply is known to survive.

in-breeding received from a Mr Wright, which could be forwarded if Darwin was at all interested.

I have no idea if Darwin even responded to this note, but I find it fascinating. It's not just that George Cupples could write to this most exalted of scientific men on somewhat equal terms (writing to Darwin these days is as close as some can imagine coming to talking to God, after all), but that there was a clear expectation that he might just be able to convey information that would be of use. Darwin's great scientific insights, which still shape so much of our thought today, rested on careful observation. Many of those observations were his own, made not least during his famous tour on the *Beagle*, but the rest were distributed throughout the world and through his crowd of correspondents.

Darwin clearly valued input from his circle of contacts. It took me a matter of moments to discover a letter, written more than 130 years ago, in which Darwin writes to Philip Sclater, an early ornithologist. Darwin was writing to thank him for a correction to some published work or another: 'You men who do only or chiefly original work', says Darwin, 'have an immense advantage over compilers like myself, as you can know what to trust'. I could scarcely have wished for a better statement as to why you want to keep those who are the source of your data close to you.

This way of working didn't end with Darwin. Strikingly similar examples of this pattern of distributed observation, reported to a central authority, exist today. My favourite recent example is the discovery of the 'ghost slug', a new species identified by staff at the Museum of Wales following reports from observant gardeners. A spectral and slimy figure, it owes its name to its nearly transparent appearance, and it can be distinguished from other, similar species by its eyeless eye-stalks. It was reported in 2008 as a new species in a Cardiff garden, and has since shown up across South Wales and also—thrillingly for me, though I've yet to find one—

as far east as Oxfordshire.* It owes its obscurity to its habit of living a solitary life up to a metre below the ground, but nonetheless someone spotted it and reported it to the museum, where it gained its scientific name (*Selenochlamys ysbryda*—ysbryd is a Welsh spirit or ghost) and a host of attention. Experts reckon that the most likely origin for this fabulous creature is not Cardiff, but the Crimea, of all places, though nothing of the sort has ever been found there. It is only thanks to observant citizen scientists out working in their Welsh gardens has it come to scientific attention.

Today, just as in the nineteenth century, transmitting information from an observer in the field to an established authority is the key to discoveries like this. Yet the relationship between Darwin and his correspondents could be a rather uncomfortable one; a letter from 16 April 1856 sees even as exalted a personage as Baronet Charles Bunbury apologizing for not having written back to Darwin sooner. He had, he explains, been waiting 'rather vainly' for 'some remark worth sending' to turn up. The point, I suggest, of writing to Darwin was that it might mean something—that Baronet Bunbury and the rest of Darwin's correspondents wanted to be of use, but that need to be useful puts a huge amount of pressure on a letter writer. Precisely because of the shared understanding that such correspondence might be useful, there was pressure to write only things which were 'worth sending'.

This paradoxical pattern, in which belief in the potential usefulness of one's contribution changes how one views a task, is something that still exists in many modern citizen science projects, including Galaxy Zoo. It lies too at the heart of why

* Observations are coordinated by the Conchological Society of Great Britain and Ireland, whose aim is to 'understand, identify, record and conserve molluscs.' I wish them all the luck in the world.

participation means more than mere crowdsourcing, more than just trying to get work done. Imagine the response of someone in rural Devon or Scotland, isolated from the mainstream scientific community and establishment, receiving a reply from Darwin praising their work, or the feeling such amateur—citizen—scientists would have had seeing their name in the (expensively printed) journals that keep track of the efforts made by observers. This is a way to understand that participation like this is a way of transforming how people think of themselves, and of their capabilities, and even in the nineteenth century it was clear, to some at least, that in asking for observations you acquired obligations to those who were assisting you to ensure that they got something from the project too.

The example I have in mind involves the Prussian/German astronomer Friedrich Argelander. He was one of the nineteenth century's pre-eminent observers of the stars, as well as a fine institution-builder and networker. Following time as a graduate student when he studied with the great mathematician Bessel, in Königsberg, Argelander moved to Finland in 1823 to head up astronomical research there. While there, he showed his dedication to observing while watching the city of Turku, home to his observatory, disappear in an enormous conflagration. The event is recorded in his log, along with a clear sense of priority: 'Here the observations were interrupted by a terrible fire, which reduced the entire city to ashes. The observatory was, thank God, spared.'

In the aftermath of the fire he moved the observatory to Helsinki, but soon after ended up in Bonn, where he had persuaded the king to fund the construction of a new, state-of-the-art observatory. (It helped in arguing for his grandiose and expensive plans that Argelander had taken care to befriend the then prince when they were still children; a serious investment in the future that perhaps modern astronomers should note.) At the time of Argelander's move to Bonn, the science he was engaged in was undergoing a revolution; it was on the edge of completing the transformation from astronomy—the measurement of the positions and the movements of celestial bodies—to astrophysics, the attempt to understand them. Argelander was interested in both, but was essentially a traditional observer.

For a long while, his most famous discovery was what was called Argelander's star. An apparently innocuous yellow dwarf, he found it moved across the sky (relative to the other stars) faster than any other star known at that time. Astronomers call this relative motion 'proper motion'; Argelander's star is still notably speedy but is now third in the rankings. Its motion is not rapid by everyday standards, amounting to a degree across the sky every millennium or so, but it is interesting. It is due partly to the star's proximity, less than thirty light years from us, but is so high mainly because the star belongs not to the rotating stellar disc that houses the vast majority of the Milky Way's stars including the Sun, but to the scattered halo of stars we now know surrounds it. As we turn with our neighbours about the galactic centre, Argelander's star stands still but, secure in the illusion that we are standing still, we conclude it is speeding by, just as the platform appears to be moving as your train pulls out of the station.

Stars don't just change position; as we saw with Leavitt's Cepheids they can change in brightness too. Only a handful of such 'variable stars' were known before Argelander began work, but he introduced the modern system of categorizing them and understood quickly that watching how a star changes is key to understanding the physics that underlies its behaviour. The only problem is that this kind of observation is immensely time-consuming. If you don't know when a star might behave in an interesting way, or on what timescales interesting behaviour is likely to occur, then you are left with no alternative but to monitor the sky as frequently as possible. Furthermore, this work was carried out star by star, and was as a result best shared by a network of widely distributed observers, a point not lost on the new director of astronomy in Bonn, who wrote:

> 'Could we be aided in this matter by the cooperation of a goodly number of amateurs, 'we would perhaps in a few years be able to discover laws in these apparent irregularities, and then in a short time accomplish more than in all the 60 years which have passed since their discovery. I have one request, which is this, that the observations shall be made known each year. Observations buried in a desk are no observations. Should they be entrusted to me for reduction, or even publication, I will undertake it with joy and thanks, and will also answer all questions with care and with the greatest pleasure.'*

It is a fabulous call to arms—'observations buried in a desk are no observations' would be a great motto for some society or other.† I love the sense of a deal being struck between those taking the observations and Argelander himself. On the one hand, we have the (presumably unfunded) volunteer with their telescope. On the other, an eminent professional scientist. Data can be passed from the former to the latter—but only if Argelander too puts his back into it and makes use of the data.

* Translation by Annie Jump Cannon in *Popular Astronomy*, 1912, from an original in the *Astronomisches Jahrbuch* of 1844.

† To my mind, greatly preferably to the Royal Astronomical Society's motto, adopted from Herschel: 'quicquid nitet notandum', or 'whatever shines, let it be observed'. Science teaches us that the real work is only beginning when observations are written down. The American Astronomical Society has a mission statement, not a motto.

Oh, and part of the deal is that he has to communicate results and answer questions from his observers.

It seems important that Argelander is offering more than a one-way exchange. As far as I know this is the first example of a professional scientist so explicitly writing about the give and take of this way of collaborating to get science done. As Galaxy Zoo took off, I certainly felt the obligation to try and respond to questions, though I can't claim to have always faced the task 'with the greatest of pleasure'. What is also reflected in Argelander's work is a somewhat formal division of responsibility; observation can be safely distributed, but analysis is specialized and central. One can argue about which is primary (and whether Darwin was being deliberately or falsely modest when referring to himself as a mere 'compiler'), but there is a settled order here.

This way of organizing things was effective, and it enabled Argelander to work on a scale that was inaccessible to astronomers of previous generations. The catalogue Argelander and colleagues put together contained the details of more than 300,000 stars, and was the definitive work of pre-photographic stellar astronomy, at least for the northern hemisphere. It remained in use for years, and his categorization of variable stars remains the standard today. If you visit the astronomy facilities in Bonn, you'll find that in 2006 they were renamed the 'Argelander Institute' in his honour; a recognition, I'd like to think, of the power of asking for help.

Networks of amateur astronomers survive too. Data on stellar variability, especially on timescales of decades or more, depend on the catalogues assembled by the American Association of Variable Star Observers, an organization with worldwide reach whose observers have assembled more than twenty million records since its founding in 1911. Rainfall observers may not, these days, form extensive networks but the Audubon Society's

Christmas Bird Count is still going strong. This annual bird-watching festival has been operating since 1900, having been introduced partly for scientific interest and partly as an alternative to the then common tradition of marking the holiday with competitive hunting. In the UK, biological recording of the presence or absence of species depends on a network of societies, many of them dating from the late nineteenth or early twentieth century, with specialisms ranging from orchids to the British Pteridological Society (ferns, since you ask).

Twenty-first-century researchers are fond of pontificating about the problems caused by big data, the sudden flood of digital information among which we struggle to pick out signals of interest. Yet the appearance of modern instruments and vast networks of the kind described above caused an earlier deluge of data, and brought a very different set of people into the scientific enterprise. These were the first 'computers', people rather than machines, and they soon accounted for the majority of staff employed at observatories.

The job of 'computer' was established at the Royal Observatory in Greenwich, for example, as early as 1836, and survived until 1937, a little more than a century. Their arrival broke the tradition by which it was the Astronomer Royal and his specialist assistants who did the work to make their own observations useful, and the staff quickly grew. The original computers, working eight hours or more a day at tedious and repetitive calculations, were recruited by looking for poor but bright students from local schools. However, as the century wore on it became clear that this was nothing more than a stopgap solution; the wages were abysmal and the work tedious, and with little prospect of promotion most computers moved on. By 1890 the Greenwich staff had hit on the idea of solving this by employing women who had university experience; such staff would be skilled enough to work as more than a mere calculator, but would lack other opportunities for scientific participation.

The papers which record the decision to employ women explicitly make clear that it was felt that Greenwich could take advantage, attracting to this often menial job women whose scientific opportunities would otherwise be lacking. If that was the marketing scheme, it was not a success. Though women continued to be employed at Greenwich, opportunities for promotion were gradually opened up for men, and those with qualifications (most often a degree) began to take up what had once been junior and menial posts, seeing them as a stepping stone to higher things. The women were once again squeezed out of even this small foothold in the scientific enterprise.

For a time, though, the position of these functionaries allowed a different sort of engagement with scientific data. What sorts of jobs were these human computers undertaking? Most of the work at Greenwich was positional astronomy, and results would be recorded in the form of measurements which were straight from the telescope, perhaps as a distance between two stars. These would have to be converted to some standard reference frame, and celestial coordinates assigned. Systematic effects like the influence of the Earths' atmosphere, which varies with the height of a source in the sky, must be accounted for. Even once that's done, single observations of a typical star are hardly going to carry much information, and catalogues must be compiled and cross checked, and global properties derived.

These calculations are the very stuff of which science is made, but just as with Argelander and his observers we see in the existence of the computers a division of responsibility. Observers—whether employees or volunteers—provide data. Computers do the processing, turning tables of data into results; the two are even separated by time, with observers producing data during the night that can be processed by computers during the day, before being studied by scientists who publish their results. Each subsequent stage of analysis depends on the former—indeed, it could hardly exist without it—but only the later stages are visible to the wider world. We celebrate the scientist who interprets the observation, not those who made it possible.

None of this is news, at least not to the accomplished historian or sociologist of science. You, my sophisticated reader, don't need me to tell that the real story of how science has proceeded over the centuries is more complex than the standard procession of dead, white, bearded men with theories might suggest, and this chapter hasn't tried to do more than offer a potted set of anecdotes. These stories do, however, illustrate that right back at the time when our modern notion of what it meant to be a scientist was being established—when we had a much more fluid idea of what science was than has been the case for most of the last century—it is possible to trace disputes about status, about the correct division of work between the classes of those involved in research. Back in the twenty-first century, we set up Galaxy Zoo to get work done. It soon became apparent that the real power and interest of the project lay in thinking about precisely these issues, and that began with volunteers doing more than just clicking on buttons to classify galaxies.

# 4

# INTO THE ZOONIVERSE

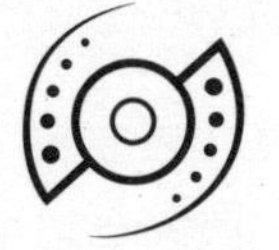

Looking back at the early days of Galaxy Zoo with more than a decade's perspective, it seems to me to be a strange and marvellous thing, this idea that so many people would give up their time to collectively contribute to science. Occasional critics carp that classifying a few galaxies isn't participating in science—that the claim to have done science should be reserved for those who designed and set up the project, and who interpret the results.

As I said in the last chapter, I'm less dogmatic. Any scientific project rests on the contributions of many people, whether it's those who operate the telescope up on a lonely mountain top or people like me, whose daily life is much more likely to involve emails and admin than a 'Eureka' moment. I know the Galaxy Zoo crowd have done science, because there's an ever-growing pile of academic papers with new scientific results within that wouldn't have existed without them.

Better still, the ideas in those papers have been adopted and echoed by the rest of the community. We were even thrilled when people started to use our results without pausing in their texts to dwell on 'citizen science', taking it as a sure sign that we were producing data of a high-enough quality that authors didn't feel the need to justify or explain their use of it. (Things took a slightly odder turn when two philosophers wrote a paper which, while calling for 'a sociotechnological turn in the philosophy of science', which I'm afraid to say you'll have to read about elsewhere, compared the rate at which Zooniverse papers were cited to others using the same data. Apparently we're as a group as productive as a world-leading research institute. Nice to know!)

The downside is that, all this time later, it's rather difficult to briefly summarize what we've found. Galaxy formation is messy, and that messiness—the fact that many different things control how galaxies first form and then change over billions of years—makes a nice, clean story hard to find, at least for now. That's what science is like sometimes, even if it makes writing a book chapter harder. So, instead of trying to present a comprehensive view, let me tell you a couple of stories that will give you an idea of the kind of thing we've been able to do with the results from Galaxy Zoo.

One problem we've tried to attack is to try and understand what happens when two galaxies collide with each other. Merging like this certainly seems important. The early Universe was filled with scrawny protogalaxies, each less than a hundredth the size of a typical galaxy today, and these seem to have, over the long span of cosmic history, gradually collided and merged to form larger and larger systems. This process isn't finished yet—the grand collision of the Milky Way with Andromeda that I mentioned earlier isn't due to happen for another four or five billion years' time, but when it does happen, our computer simulations make it clear that it's likely to be a spectacularly messy and disruptive event.

When two large galaxies like these collide with each other, a cosmic ballet ensues. The first approach sees the galaxies fly past each other, their mutual gravitational attraction distorting their previously neat discs and creating long streams of stars (Plate 6). These are tidal tails, unstable creations of the merging process, and as they begin to fall back towards the main body of each galaxy the two discs turn and plunge back together once more.

This repeated encounter creates new distortions, and further disruption as the merging system takes on a wide variety of forms. We can see this stage of the process in nearby galaxies, whose names conjure up appropriate images—the 'Antennae', with two long streams stretching away from a bifurcated body the 'Mice', imaged beautifully with the *Hubble Space Telescope*, with long tails revealing a recent interaction. Such a stream has even been spotted stretching between Andromeda itself and the third large member of our Local Group, M33.

Apart from being flung out of orbit, stars which formed before the merger will continue as they were before; even within a galaxy of a hundred billion stars like the Milky Way, there is enough space in space to make a collision between two stars during a merger vanishingly unlikely.

That's not to say we shouldn't expect fireworks when the Milky Way and Andromeda collide. Gas clouds do collide with each other and the result is a spectacular boom of star formation. The Earth may not be the best place to watch, as the Sun will by then have entered its red giant phase and swallowed our home,* but if you can make it to a suitable planet then you should expect a spectacular night sky, speckled with newly formed and brilliant,

* I may be being unfair to the Earth's prospects as a long-term observing platform. As the Sun converts hydrogen to helium it loses mass, and, because of the law of conservation of momentum, our planet spirals slightly outwards. There is therefore some chance that the Earth may survive the Sun's swelling into a red giant, though how reassuring you find the chance of our planet's future existence as a charred cinder is perhaps a matter of personal taste.

massive stars. The view from inside what's called a starburst galaxy must be absolutely wonderful.*

Though such a spectacular rate of star formation most likely can't be sustained, the long-term effect of a merger might be to change the shape of the colliding galaxies for ever. In the case of the collision between the Milky Way and Andromeda—two discs—the result according to most simulations is their transformation into an elliptical.

This makes a certain amount of intuitive sense; discs are ordered systems, their stars orbiting in concert around their centre, and a serious disruption will see stars kicked up out of the disc and into the more random pattern of movement which characterizes ellipticals. A new, unified galaxy is produced (what some researchers insist, despite everything, on calling Milkomeda or—hardly better—Milkdromeda), larger and more massive than before and ready to continue life as a stereotypical elliptical.

Perhaps the last stage of such an event takes place deep inside the new galaxy, at its core, as the supermassive black holes that previously inhabited the centres of the constituent galaxies dance slowly around each other, losing energy in the form of gravitational waves and spiralling inwards, eventually merging. A small number of galaxies are known that have double or even triple black holes at their centres; though they look otherwise undisturbed, these are most likely the products of recent mergers.

Galaxy Zoo must contain many such galaxies, observed a few billion years after the end of the merger. Can we tell, just by looking at the galaxy, that anything spectacular had happened?

* Of course, the odds of your planet being blasted with lethal radiation from a nearby supernova is greatly increased in such a system. One can't have everything.

The tidal tails of stars flung from the centre of the system will eventually disperse, their stars lost to intergalactic space or once more part of the main galaxy. There's not much hope in finding definitive proof of a merger just from shape.*

Nor is the colour of a galaxy likely to be much good. The colour tells you what's happening right now, and the great burst of star formation that accompanies a really dramatic merger most likely lasts only a few hundred million years. Galaxies which have survived mergers will not easily be distinguished from those which have evolved through less dramatic means, which makes it difficult to try and understand the effect of mergers on galaxies. Such events look spectacular, but we'd like to understand how significant they really are.

For example, it's possible that most stars form during the dramatic starbursts that follow a collision. It's also possible that, like a firework display, these collisions are spectacular but ultimately have little long-term effect; in this scenario, most stars form because of other processes, and galaxies would look much the same in a universe where mergers were much less likely.

How can we distinguish between these possibilities? What we really need is the chance to do a direct experiment. I'd love to assemble a vast, intergalactic laboratory (along with, of course, a few billion years' worth of funding). In it, I would assemble two populations of galaxies. The two populations would begin the experiment in identical states, but in one tank gravity would allow the galaxies to merge, just as happens in

* You might just be able to do something by looking at the outskirts of the galaxy and hoping to detect the faint leftover scattered debris, which can, in some cases, persist for billions of years. New instruments, using special lenses first developed for photographing high-speed motorsport of all things, are useful, but such observations are time-consuming and as yet have only produced data for a small number of systems.

the real Universe. In the other, an ever-watchful PhD student could be tasked with intervening to keep the galaxies apart from each other. (I have no idea how they would do this, but as we're imagining a laboratory in which there are two tanks at least a few hundred billion light years across, I think we're allowed a little magic.) After ten billion years or so, we could then compare the resulting mix of galaxies in each tank, and see what effect merging really produced.

Of course, such an experiment is impossible for several very good reasons. We can attempt something similar in a supercomputer—and people do—but it is a tricky problem. Keeping track of the galaxy-scale details of the merger and the small-scale processes of star formation that determine how the galaxies look at the same time is a difficult computational challenge, and requires shortcuts to be made.

As a result, while some might be satisfied with the results of simulations, I prefer to look out into the Universe for my experiments. Somehow we need to assemble a tankful of galaxies that have managed to avoid merging. These will be rare, but they are out there, and they reveal their presence via their shape—exactly the kind of thing that the Galaxy Zoo volunteers can help with.

I often remember Patrick Moore's claim that the Milky Way resembled nothing more than two fried eggs, 'clapped back to back', with a thin disc surrounding a central bulge, represented by the eggs' yolks. This gives you a good way to think about spiral galaxies,* and central bulges are so common that at least one astronomer has proposed that both elliptical and spiral galaxies should really all be seen as nothing more than bulges, some of which happen to have grown discs.

* It is, however, a lousy way to serve eggs.

One way of producing bulges is via a galaxy merger, as the disruption kicks stars out of the main disc and into the bulge. Even a small collision should add substantially to a bulge, which is why it was surprising when, among the hundreds of thousands of galaxies searched by Galaxy Zoo a small number of unusual objects started to appear in the classifications. These special systems have no visible bulge, and are therefore guaranteed to be merger-free. We can put them in our second tank, and compare them to galaxies with more normal histories.

Nature and our volunteers had provided a way of doing the experiment I imagined above. In charge of the experiment was Brooke Simmons, a Californian who ended up working with me in Oxford before winning a prestigious Einstein Fellowship from NASA. She took the latter back to California, craving decent Mexican food and more sunlight than Oxford could provide (our department provided a special lamp for her to cope with the winter which replicated the exact spectrum of natural sunlight; it was apparently cruel otherwise to keep an American in English conditions*).

Brooke's speciality is understanding how the supermassive black holes at the centre of galaxies grow, and so our first experiment was to try and see if mergers contribute significantly to their growth. Most people expected that they would; even galaxies which have experienced many mergers have a single, massive black hole at the centre, not a whole cluster of little ones, so merging must happen.

Further evidence that galaxies' central black holes grow through mergers comes from the best-known bulgeless galaxy, NGC 4395, which had been studied a decade earlier and was

* It's amazing to see how many people who scoffed at this provision now gather around its light in the winter months, myself included.

found to have a puny black hole, more than ten times less massive than we would otherwise expect for a galaxy of its size. No mergers, no black hole growth, it seemed. But that result is only from one system. Now, with many more merger-free systems to play with, we could do the experiment properly.

That means measuring the mass of each galaxy, and then the mass of the central black hole. Estimating the mass of a galaxy isn't too difficult. We know how a star's colour and brightness depend on its mass, and a galaxy is just an assemblage of stars. (Yes, there's dark matter and dust and other things, but we're concerned with the stellar mass here; for most large galaxies the ratio of total mass to stellar mass is pretty constant.) The brightness of the galaxy tells you how many stars are there, and the colour tells you what sort of stars they are. Know those two things, and you can get a pretty good handle on the mass of the galaxy.

The black hole mass is trickier. Measuring the mass of something that's invisible, and which is in any case tiny on galactic scales, isn't easy, but we can get there via an indirect route. What we do is look right at the centre of the galaxy, expecting to see the bright glow of hot material as it falls down into the black hole. Such accretion activity is most easily seen in the x-ray region of the spectrum, but in the visible what you see is a star-like point of light. For each galaxy, Brooke took all the light that could possibly belong to such a source, and assumed that it belonged to the material falling into the black hole.

That gives us a guess at the rate at which the black hole is growing. It's a start, but we want to know how massive the black hole itself is. Luckily, how massive a black hole is turns out to be tied to the rate at which it consumes material. The idea that black holes do anything but voraciously devour everything around them might be surprising; it certainly belies the fearsome reputation they have in science fiction, where they're always 'lurking' at the centre of a galaxy, rather than just hanging out in space minding their own business.*

This has always seemed unfair to me, and I'm almost tempted to found the Society for the Promotion of Friendly Black Holes. Our promotional material will make much of the careful manners these exotic beasts exhibit. It turns out there is a maximum rate at which, under normal circumstances, black holes will consume fuel. This slightly surprising result is due to the dramatic effect on material that, falling into the black hole has. It heats up and shines brightly. The presence of this radiation creates a pressure, pushing outwards and preventing more material from falling in; the whole thing becomes a self-regulating system with a maximum rate of accretion known as the Eddington limit.

Using this piece of information, we can convert the minimum observed rate at which material is falling into the black hole into an estimated mass, and finally make the comparison we want to.

I've probably given too much detail here, though there are many more gory specifics I could have included (we have, for example, only really calculated an estimate of the maximum mass). I do, however, want to make the point that this is the meat and potatoes of modern astrophysical research. First, we carefully assembled a sample of interesting, distinct galaxies. Once they were found, we selected a comparison sample, and then we measured their properties, being careful to make sure we had as much understanding as possible of what the likely errors were

* A star in a distant orbit around a black hole is in no more danger of being sucked into its cavernous maw than the Earth is of falling into the Sun. It's true that if you get too close, you'll fall inevitably into the black hole itself, but that's hardly the black hole's fault. Star trekking travellers of the future need not worry about being consumed by a black hole, they need simply to study astronavigation properly.

and how they'd affect the result. It's painstaking work; each of the images of each of the galaxies had to be calibrated by hand, and the whole effort probably took Brooke something like six months.

That careful work paid off, though, and these weird galaxies now told us something about the Universe and its history that we didn't know before. It turns out that the bulgeless galaxies have black holes which are pretty much the mass one would expect from a normal galaxy. In other words, despite never having ever had a significant merger, living instead lives of splendid isolation, they manage to grow large supermassive black holes.

This research isn't finished. I've included in this book a brand new image of one of our bulgeless systems from the *Hubble Space Telescope* (Figure 17), which is allowing us to measure its properties with new accuracy, and thanks to help from some friends we've been able to find merger-free galaxies in one of the big

**Figure 17** A bulgeless spiral galaxy, discovered by Galaxy Zoo volunteers and observed by the *Hubble Space Telescope*. *Hubble*'s sharp resolution allows us to peer into the heart of the galaxy.

supercomputer simulations and to compare the properties of such systems in this artificial Universe to those in the real one.

If those simulations are reliable, they confirm a hard truth that astronomers have faced for years. I often end my talks with a bleak version of the far future of our Universe, destined (we think) to become an nearly empty void, a vast sea of space expanding forever into yet more nothingness.* What seems to pack a greater emotional punch, though, is this: the Universe is already past its best.

More stars are dying each year than are being born and galaxies are shutting down in a process that's been going on for billions of years, since the youthful, exuberant peak of star formation that took place shortly† after the Big Bang.

What's causing this cosmic shutdown? There are lots of ideas around, starting with the simple idea that we're out of fuel. Stars need cold gas to form, and in some galaxies at least the reservoir seems to have been exhausted. Maybe galaxies normally rely on a flow of material from outside to keep the stellar factory going, and that's disrupted by close encounters with other galaxies. Maybe it's falling into a large cluster of galaxies that triggers that process. Maybe a collision between two galaxies causes a burst of star formation so dramatic that it uses up all of the available gas. Or maybe if a large galaxy like the Milky Way consumes too many smaller systems then the effect is the same. Or maybe we need to look back at the central black holes; the complex physics and twisted magnetic fields that exist around them can fire jets of material moving nearly at the speed of light out into the galaxy, heating or expelling the gas it encounters. Maybe if you form too

* I do like to send an audience home happy.

† Shortly here means a few billion years; astronomical timescales are hard to get used to.

many massive stars the resulting burst of violent supernovae which marks their end can similarly heat the surrounding gas.

And on and on and on. Understanding exactly what's going on is a difficult task, but it might be important in understanding our own Milky Way. Galaxies which are still vigorously making stars are, on average, bright blue, lit up by their new creations. Those where star formation has ceased—we would say 'quenched'—are red. The Milky Way, at least according to work by a bunch of Australian researchers, is neither red nor blue, but green, which means that it seems to be undergoing this transition right now. If they're right, then we happen to catch our galaxy at an unusual point in its history and must look outward to understand what's happening.

Trying to distinguish between so many possible causes makes carrying out a clever experiment like the one we managed with the bulgeless galaxies near impossible. Instead, we use the sheer scale of the Sloan survey and the classifications provided by Galaxy Zoo, pile them all together, and look at which galaxies have which properties.

This task fell to my first PhD student,* Becky Smethurst, now an independent research fellow at Christ Church here in Oxford. Before she arrived, I was terrified of taking on the responsibility of supervising a PhD student. Your PhD supervisor sets the direction and tone of your research, so while the student is ultimately responsible it seemed very easy for me to completely ruin someone's career. Luckily (apart from an unfortunate incident which resulted in American Express bombarding me with advertising for Taylor Swift concerts, the less said about which the better) Becky and I got on well and she proved more than smart enough to deal with my blunders.

* Oxford insists on awarding not a PhD, but a DPhil, thus leaving everyone involved explaining their degree for the rest of their career.

Her task wasn't easy, and Becky constructed a sophisticated apparatus of modern statistics and computer-based analysis to look at this problem of quenching. The result of all that effort? Well, to almost no one's surprise it turned out to be complicated. Different galaxies seem to go through the transition from blue to red in different ways. Some, mostly elliptical galaxies, seem to have shut down their star formation rapidly, perhaps the result of a spectacular merger. Others, including most spiral systems and the Milky Way itself, quench more slowly. Some systems that sustain growing black holes show clear signs of recent and dramatic quenching, so it's clear that they can play a role too.

Where a galaxy lives also makes a difference, with those that find themselves in more crowded environments undergoing a more dramatic shutdown than their relatively isolated cousins. In other words, we shouldn't carry round a picture of a galaxy as an isolated system (the 'island universe' that a galaxy was once thought to be) but we should rather think of them as interacting with their surroundings. Galaxies at the centre of a large cluster—the nearby Virgo Cluster, say, which contains more than a thousand galaxies and which weighs in at more than a million billion solar masses—have a very different life from those living in more rarefied parts of the Universe.

These results—Becky's work on quenching and that led by Brooke on bulgeless galaxies—are just two examples of the things that are made possible by the careful classifications provided by Galaxy Zoo volunteers. It's been incredibly satisfying to watch the project team use our volunteers' efforts to understand more about the Universe, and to see the use to which other people have put them.

I have, though, continually been distracted while the frenzy of astrophysical research unleashed by the availability of the Galaxy Zoo results was playing out. Almost as soon as the project started

I began to get phone calls and emails from other scientists, in fields about as far removed from my own as is possible to imagine, who wondered whether our Galaxy Zoo volunteers might be willing to help them out, too. The enquiries ranged from the polite to the pleading, but they revealed what I should have known already. Astronomers were not the only researchers struggling with the sheer volume of data now accessible to them; whether ancient historian or zoologist, they were likely suffering from the same set of problems, and citizen science, Galaxy Zoo-style, seemed like a way out.

By this point—somewhere in 2008, a year after the launch of Galaxy Zoo—I'd abandoned any pretence of doing the work Oxford had employed me for in the first place, and was thinking about citizen science full-time. We were given a small amount of money by Microsoft, part of a fund set up by the company to commemorate their leading computer scientist, Jim Gray, who had vanished while sailing near San Francisco in early 2007. (Our project appealed partly because of the response to Gray's disappearance; his colleagues and friends organized a distributed search for signs of his boat in satellite images, a task which was eerily reminiscent of the kind of thing we'd ask Galaxy Zoo volunteers to do in the years ahead.) A grant from the wonderful Leverhulme Trust followed (I love an application form which includes the question 'why won't anyone else fund this?'), and for the first time we could think about expanding.

From this point on 'we' includes a diverse cast of wonderful, slightly bonkers web developers and escaping scientists who deserve a lot of credit for the last ten years. This isn't a formal history, so I won't stop along the way to describe who did what, but you should be very aware that this is a team effort. My first hire was Arfon Smith, a cheerful Welsh presence who I'd first encountered when we were PhD students. Arfon studied astrochemistry from a chemist's perspective, but had realized research was not for him and become a web developer. (He now heads the grandly titled Data Science Mission Office at the Space Telescope Science Institute in Baltimore, the home of *Hubble*, and so his attempt to escape academic research hasn't gone brilliantly.)

With Arfon's expertise on board we could build something a bit less jury-rigged to support Galaxy Zoo, and also try out some new things. We were planning the Zooniverse—a platform consisting of many projects—rather than just a single website. As we thought hard about how to build a system that could be what we wanted, we slowly acquired a set of test projects, all ambitiously different from each other. One of the first came when I found out that the team at the Royal Observatory Greenwich were thinking of developing a similar project using solar data. I've always loved visiting Greenwich—arriving there on the boat which runs from central London and spying the green dome of the old observatory up on the hill, behind the beautiful old Royal Naval College and Queen's House, is absolutely thrilling. I was also intrigued by the idea of working within a museum; the success of Galaxy Zoo had meant we were suddenly communicating with a huge number of people in a fairly novel way, and the idea of a place where there were experts in communicating with the public seemed useful. More to the point, as far as I was concerned if anyone was going to do such a project it was going to be us.

I recruited Chris Scott (then at the Rutherford Appleton Lab, now at the University of Reading), who I'd interviewed several times on the topic of solar weather. Chris is one of the team behind a very special pair of cameras, the Heliospheric Imagers (HIs) on the twin *STEREO* spacecraft. *STEREO*'s mission was to study solar weather, the activity on the surface of our star which can affect the whole Solar System. At any given time, particles are flowing away from the Sun in a stream known as the solar wind, but occasionally things get more spectacular.

The Sun is a ball of ionized gas (otherwise known as plasma). That means that as it rotates, it doesn't do so in the way a solid body does. If you could stand at the solar equator (a terrible idea for many reasons), you would complete one rotation every 24.5 Earth days. If you stood at the solar poles—a no less terrible idea—it would take thirty-eight days to rotate. The Sun also has a strong magnetic field, and this differential rotation has a profound effect on it. The magnetic field becomes tangled, and every so often releases energy by springing back to an untangled form, expelling material into space as it does so.*

These events are known as coronal mass ejections, or CMEs for short (Plate 7). In Chris' phrasing, each consists of a billion tons of matter moving at about a million miles an hour. They are spectacular and dramatic, but they are interesting for practical reasons too. Every so often, the Earth happens to get in the way of one of these CMEs. If conditions are right (again, the details depend on the complexities of interacting magnetic fields and charged particles), the particles from the CME can cause a change in the Earth's upper atmosphere, creating glorious displays of what are called the aurorae, the Northern and Southern Lights.

The background flow of particles from the Sun means that at least a faint display of aurora is visible on most nights. I used to act as a tour guide on special flights to go view the Northern

* The actual physics of this are, from my perspective, unbelievably complicated, and as a result I have the utmost respect for the scientists brave enough to take on trying to understand the Sun as their life's work. If you want to bamboozle most astronomers, just ask if they have considered the effect of magnetic fields; the answer is almost always 'no'. I have the liberty in my work of looking at distant stars and galaxies and deciding that they look simple; with the Sun, we have no such option and must confront its complexities.

Lights; we'd fly north towards Iceland, turn the lights in the plane off (and sometimes those on the plane's wings too), and peer out of the windows. I took something like forty flights, and only once did we have a complete failure. Most of the time, though, what we could see was a faint, grey curtain. It might flicker a little, and change shape over the course of the hour or so we'd watch, but my job as the on-board astronomer was to make sure people were excited by seeing something so unspectacular.

After all, most of the people on board the flights were there because they'd seen footage or photos of brilliant and brightly coloured aurorae, lighting up a snow-covered landscape with red and green shadows as the lights dance overhead. I've seen such a display only once, on a trip to Tromsø in northern Norway. Local aurora expert Kjetil Skogli had taken our group out to a frozen lake, but as we headed to the site we could already see something spectacular was happening.

Before too long, I was lying on my back in the snow looking up into a clear night sky that was like nothing I'd ever seen before. The horizon was lit up with a bright green curtain that seemed to change ethereally even while we looked at it. Bright streamers reached up, high into the sky, suddenly brightening and fading as I looked. Eventually, the sky far above me was encircled with red, a feature known as an auroral crown (Plate 8) which I'd only read about before. It was utterly magical, a transformative hour or so that I will never forget, a few moments' glory powered by the arrival of a CME, with particles from one of these events pouring down onto the Earth's atmosphere and exciting the particles there to glow brightly for our entertainment.

The next night we went back out, and drove through a blizzard to the Finnish border to find a gap in the clouds. Despite this, we were rewarded only with a faint glow. The Sun, and its interaction with the Earth, is capricious in the extreme. Yet particularly large or energetic CMEs can have consequences that reach beyond the success of a sightseeing trip. The electric currents induced by such activity are a serious threat to much of our electronic infrastructure; a power blackout that affected large parts of North America and Canada in 1989 was blamed on a solar storm. The famous Carrington Event, a dramatic flare observed during the nineteenth century, affected telegraph systems—the high-tech communications infrastructure of the time. Much of our infrastructure is now in space, with satellites in the firing line and vulnerable to the effect of CMEs.

With warning, most of these negative consequences can be prevented. Understanding solar weather, and predicting whether an observed event might hit the Earth, has thus become a priority. To this end, the *STEREO* mission was supposed to provide a unique perspective. It consisted of two separate spacecraft, each placed on an orbit which meant it drifted slowly away from the Earth. One was ahead of our planet and one behind, flying with cameras turned to study the Sun and its environment.

The HI cameras the twin spacecraft carried had a different job. They were designed with a series of internal baffles, made of some of the blackest material available, all in the service of reducing internal reflections. That's necessary, because these cameras had the job of staring at the space between the Sun and the Earth, watching for the faint trace of coronal mass ejections travelling through space.

The images from the HI, turned into low-resolution video, are strangely beautiful. You see nothing but a background starfield at first, drifting slowly past the camera as the spacecraft moves. You might notice a couple of stars that are much more brilliant than the others. These aren't stars, but planets—Venus, or even the Earth drifting slowly through the field of view. The fact that we can launch a spacecraft capable of capturing a beautiful image of our own planet, as just one drifting object among a myriad stars, is something that stops me dead in my tracks from time to time, but this—and the occasional spectacular movie of a comet having its tail removed by the fast-moving particles of the solar wind—is very much beside the point. The main goal for the *STEREO* HI imagers, and the other cameras on board, was to understand how the solar wind gets launched and then travels through the Solar System.

Picking out the ghostly trace of a passing coronal mass ejection against a background of stars is difficult—exactly the sort of pattern recognition task that computers still struggle with and at which our biological, evolution-honed senses excel at. With the help of those in Greenwich, Chris' team, and others, we created a project, Solar Stormwatch,* which asked volunteers to watch videos, spot CMEs, and trace their progress across the Solar System.

This was a difficult task for us, as we had to build a whole new set of tools capable of dealing with video and allowing the careful marking that was needed to produce scientifically useful results from such a task. It was also difficult for the volunteers, who had to look carefully to find even the faintest traces of activity in very busy images. Of special importance were the first few frames of any particular event, when the particles that made up the CME had just been launched from near the surface of the Sun; as they travelled through the lower atmosphere of their star, they would have interacted with the magnetic fields that thread the region, producing what could sometimes be a dramatic effect.

To our delight, Solar Stormwatch was a hit. While not receiving the publicity that Galaxy Zoo had, tens of thousands of

* Most of the development was done by the talented Jim O'Donnell, then a web developer in the team at the museum in Greenwich but more recently a stalwart of the Zooniverse team in Oxford.

people took part, and there were a couple of immediate scientific results from the Solar Stormwatch project. First of all, Chris' team showed that—partly because of the unique vantage point afforded by the two STEREO spacecraft and partly because of the care taken by volunteers to achieve incredible accuracy—using classifications from citizen scientists provided a better warning of the approach of a CME towards Earth than existing automated systems. If you're a company with commercial satellites, you would literally be better off consulting our crowd (and, yes, launching your own version of *STEREO*, as the vagaries of orbital mechanics have since seen the two spacecraft drift to less useful positions) than relying on your resident machine-learning experts. Machine learning for this sort of problem is still in its infancy, and professional forecasters, it turns out to my surprise, still inspect the data by eye, but perhaps the Solar Stormwatch results might provide a gold standard set on which future storm-hunting robots could be trained.

There was also a scientifically interesting result. I described CMEs earlier as if they were the result of a sudden event, after which they just coast out into the Solar System. Instead, our data confirms what had been seen in images from the *SOHO* satellite further from the Sun's surface; the particles that make up the CME accelerate away from the surface of the Sun. Rather than just setting off into the Solar System, they get pushed on their way by the complex magnetic fields that exist close to the solar surface.

Or at least that's what seems to be going on. Unfortunately, I for one didn't realize this was going to be the interesting bit, and so we designed the Solar Stormwatch interface without paying special attention to the few frames the *STEREO* cameras capture around the point where the CME is just beginning to head out into the Solar System. We've fixed that now, and a newer version of the project has asked classifiers to pay particular attention to this most interesting of regions, but the results are still awaited eagerly.

Solar Stormwatch seems a large jump from Galaxy Zoo, but we were looking in other directions too. I knew that a large number of planetary scientists spent their days counting craters, the better to understand the history of the worlds they were studying, and the task seemed ripe for citizen scientists to sink their teeth into.

The principle behind crater counting is simple. Imagine that, at some point in the Solar System's five-billion-year history, the great volcanoes of the Martian range sputtered into life, with lava flowing out to cover the surrounding plains. The new surface would be, at least when seen from orbit, smooth and new. Wait a few million years, though, and a meteorite large enough to survive a fall through the thin Martian atmosphere is likely to hit, burying itself into the surface or vaporizing near impact, in either case leaving behind an impact crater. Wait a little longer, a second meteorite might hit. And then another. And another. Even seemingly rare events become common over the billions of years of cosmic history.

From today's perspective, we can count the craters to work out how old the surface is, at least relative to others on the same planet. The cratered surface of the Martian highlands, for example, is clearly older than the volcanic and smooth slopes of the Tharsis Montes. Play this game on the Moon, and there's an added twist: the Apollo missions brought back a treasure trove of lunar rocks.

A small proportion of this bounty was used for symbolic purposes, with fragments of rock distributed around the world as symbols of American largess and technological superiority. (Though Britain's allocated lump sits proudly in the Natural

History Museum in South Kensington, many countries have lost theirs; Ireland's, for example, was thrown away after an observatory fire damaged the building it was kept in.) Most, though, has been kept in pristine conditions for scientific purposes, among the most important of which is establishing an absolute, not a relative date.

With the lunar rock returned to Earth, geologists and planetary scientists have been able to use the full gamut of laboratory techniques to establish the age of the samples, mostly via radiometric dating. One technique makes use of the fact that a particular isotope of potassium, $^{40}$K, decays to a particular isotope of argon, $^{40}$Ar, when one of the protons in its unstable nucleus turns into a neutron.

Argon is a gas, and if the rocks are melted through heating, it escapes. Measuring the amount of $^{40}$Ar present and comparing it to the amount of potassium reveals how long it's been since the rock was melted. Since for most of the lunar surface the last melt corresponds to the volcanic formation of the surface, for the few precious places visited by the twelve Apollo astronauts we know absolutely how old the surface is. Count the craters there, and we can calibrate our entire scale for the Solar System, and get proper dates for major, surface-marking upheavals on any world.*

It still boggles my mind that our understanding, say, of how old the moons of Mars are, needs to be calibrated by work with a lump of rock picked up from the Moon by human beings, but until we explore further that's precisely the case. The trouble is the crater counting. Most of the Moon's surface is billions of

* It is, of course, much more complicated than that. To do the job properly one needs to consider the different rate of bombardment on Mars, neighbour to the asteroid belt, for example, and on the Moon, and of course the different rates at which meteorites burn up in Mars' atmosphere compared to, say, in the thicker air of Earth, will matter too. But you get the idea.

years old, and that is a lot of time for craters to build up. There are old surfaces on Earth, too, but here erosion due to weather removes the scars of all but the largest or more recent impacts. On the Moon, more or less, once a crater is in place it stays put until aeons' worth of subsequent impacts eventually obscure it.

Looking at parts of the Moon is a matter of picking craters out from among the debris of previous generations of craters, which is not an easy task. As higher-resolution images became available, the task of crater counting became more and more difficult and time-consuming; the arrival at the Moon of NASA's *Lunar Reconnaissance Orbiter*, which mapped almost the whole surface at a resolution of a hundred metres per pixel, threatened to overwhelm scientists (Figure 18).

**Figure 18** The Apollo 17 landing site as seen by the *Lunar Reconnaissance Orbiter*. The remains of the lunar module and blast marks from its take-off are clearly seen, as are tracks from the lunar rover the astronauts used to explore.

The solution was to turn to volunteers, and using the same software that powered Galaxy Zoo we set up and ran a project called Moon Zoo, producing catalogues of craters at important sites. The project wasn't as successful as its predecessors—partly because there turned out to be pretty big differences in what experts were prepared to count as a crater—but it was another early reminder that we galaxy experts were not the only ones facing a flood of images and data that people were struggling to deal with. Nor was it just astronomers. With the Sun, Moon, and the distant Universe covered, the fourth Zooniverse project brought us back down to Earth.

Back down to Earth, though still dealing with planetary-scale phenomena. Understanding the Earth's atmosphere and climate is of inherent interest even to an astronomer like me—how else are we to get a sensible understanding of how all of these planets around other stars must behave—but being able to predict how the climate might change in the next few decades is an urgent, vital question. As it's not possible (or at least, not sensible) for them to conduct experiments on a planet-wide scale, climate scientists have reached the same solution as cosmologists who face similar struggles experimenting on the Universe, turning to large computer simulations.

In the powerful computers owned and operated by places like the UK's Met Office, it's possible to run simulation after simulation of how the weather and climate might play out over the next few decades given different scenarios. It's a complex, difficult problem, replete with the kind of feedback loops that mean it's hard to make even simple predictions. For example, warmer air may lead to more clouds. But think about an image of the Earth from space, with bright white clouds flecked across the darker blue of the oceans and the green and brown of the land. It's not hard to realize just by looking that the clouds reflect more light than the surface they cover, so they have a cooling effect. But more clouds means more rain, and so the composition of the atmosphere changes, and so on and so on.

These difficulties can be addressed, and indeed they have been to such an extent that we can be confident that the changes now visibly underway in the Earth's climate are due to the steady input of man-made carbon dioxide into the air. Unless politicians the world over (and, yes, the rest of us) get our act together, these sobering simulations show our future of more extreme weather events, unbearable summer heat, and dramatic change across the face of the planet. Planning for these changes, much less averting them, depends on being able to rely on the models, and that means doing the best possible job of testing them.

One option is to wait. I did attend one meeting of climate scientists studying the Arctic, memorably held in a dark and stormy Reykjavik one October. With storm clouds gathering over the ocean and cutting the days short, it appeared to me to be a suitable location to discuss the fate of the world. I'd expected the speakers to have ideas on how to engage the public in averting climate catastrophe, but the mood of the meeting was different. Everyone who was involved in trying to create explanations, theories, and models of the Arctic had results that made sense of current conditions. That, after all, was the price of entry—any new theory that couldn't explain what was happening in the Arctic today would get short shrift. However, different models predict wildly different futures, and amid a slightly manic sense of end-of-the-world excitement, the scientists in Iceland discussed how the changing climate might provide experimental proof of one idea or another.

Given that all of the models predicted significant rises in sea level, I prefer to find a means of testing climate models which doesn't involve permanently ruining the delicate balance of the

Earth's atmosphere and ecosystems. It seems a bit much to precipitate the greatest environmental catastrophe in the history of our species to prove one *Nature* paper right over another. Luckily, we can test climate models by seeing how well they predict not only the future, but the past.

By gathering data on the weather and climate going back centuries we can make new demands on the supercomputers and the theories which are programmed into them. As well as explaining what we see today, we can insist they explain the past too, and thus gain confidence that they will be able to guide us towards the future. The only problem is that we don't have decent records of the weather across much of the world from before the middle of the twentieth century. Western Europe and the Atlantic coast of North America are OK, but the rest is pretty hazy. My main contact at the Met Office, a physicist called Phil Brohan, describes the picture as being clouded by a fog of uncertainty.

Climate is global, and the effort to understand how the atmosphere is behaving greatly benefits from a worldwide picture. One part of the world might be having an unseasonably rainy summer while others bake in conditions of severe drought. If Europe is cold one winter, it doesn't mean that China isn't having a mild time of it. Records do exist, but they are locked away, hidden in handwritten notebooks and stuffed in drawers in half-forgotten archives the world over.

Phil reckons there are a billion or so observations out there to find, in need of rescue and conversion into digital data that can be fed into the computers. Some have yet to be unearthed from those dusty drawers. Most still sits stubbornly on paper, awaiting the digital photography that is the first step in the long process of digitization. Some, though, just need typing up, a data entry task that seemed both urgent and so large as to be intractable.

We thought that volunteers might be able to help, but their work needed focus.

Phil and colleagues wanted to concentrate on the most useful data, something that would give them worldwide coverage as quickly as possible. You really want a fleet of mobile weather-observing platforms, touring the world and making scientific measurements as they went. The word 'fleet' is the right one—land makes things complicated, what with hills and mountains and valleys and other factors that affect the local weather, so to keep things simple you would, for preference, send your weather detectors out to sea.

Luckily, the British Royal Navy has been systematically recording the weather every four hours on board every ship since the late nineteenth century, sticking to the task come hell or literally high water. The ships' logs held in the National Archives in Kew contain page after page of these observations in table after table, with air pressure, temperature, and information about the wind recorded next to the everyday business of loadings, unloadings, and navigation. These logs aren't brilliantly written literary journals penned by officers who would, were it not for Naval service, be dashing off novels in London, but to the Met Office team they were priceless treasures, and we decided to start with the logs from the early part of the twentieth century, particularly around the First World War.

If we could get people to transcribe what was written down, that is. The software was easy enough to adapt, but there was a nagging worry that people would find the task just too boring to be contemplated, no matter the scientific justification. My confidence that we weren't just inflicting tedious data entry on people wasn't helped by the quizzical look the project name we chose got in meetings; we called the project 'Old Weather', which I thought captured nicely the everyday nature of the data being collected and the public fascination with weather records. It turns out that talking about the weather is, perhaps, a uniquely British phenomenon (who knew?) and my American colleagues in particular were deeply sceptical.

We were so worried that we did something we hadn't done in Galaxy Zoo, Solar Stormwatch, or Moon Zoo. We decided to try and turn this task into something of a game. When they arrived on the site, classifiers were expected to sign up to join a ship's crew. They could then follow along on its voyage, merrily typing in weather data as they went, and would be rewarded for their efforts by promotion within the ranks. The volunteer who'd contributed most to a particular ship's logs would be the captain, those who had contributed a bit the officers, and so on down the list to the new recruits who had merely dabbled.

It turned out that we were silly to worry, and Old Weather was an enormous success. In concentrating on the weather data our friends the climate scientists needed, we had neglected the inherent interest of the logbooks themselves. Within their pages, which we had seen only as tables of numbers, were laconic records of life on-board ship. As they worked through page after page, volunteers got to follow their ship around the world. (This turned out to be useful, too, as volunteers who spotted sudden jumps in recorded position could put right mistaken coordinates which were wrong in the original logbooks themselves.)

They also found odd little notes providing windows into life on the high seas. There were reports of illness, and even the occasional death. Comings and goings were recorded; one officer who popped up on several ships turned out to be responsible for distributing medals to boost morale. One, presumably long-suffering crew seems to have enjoyed regular lectures on a variety of improving subjects from their captain, all faithfully noted in the log. We know, thanks to Old Weather volunteers, where a saucepan was lost overboard just south of Iceland. Participants were especially moved by the loss of the chocolate rations of the HMS *Mantua* in a dockside loading accident (Figure 19).

These minutia have now been edited—by volunteers—and are available online. The weather data has been fed into climate models and the fog of ignorance Phil and his Met Office colleagues are fighting has receded just a little. Having completed the Royal Navy's First World War logs, Old Weather volunteers have taken on more difficult challenges, including the amazing records of the early Atlantic whalers whose battles with the ice preserve a record of exactly where that ice was.

And the list goes on. In 2010 I moved, temporarily, to the Adler Planetarium in Chicago. Adler's a marvellous place, sitting on a peninsula that juts out into Lake Michigan; my office was underneath the best view of the city skyline available anywhere. Founded in 1930 by Max Adler, a businessman who'd made his fortune from the Sears catalogue empire, it's nearly unique as a place which employs academic researchers as part of the museum staff. They'd spotted the potential for volunteers to contribute to science long before, and were excited about Galaxy Zoo. While I was there, a grant from the Sloan Foundation meant we could build a proper development team. I soon moved back to Oxford, but Arfon moved out to Chicago and the Planetarium has hosted a large part of the Zooniverse team ever since. Before too long we were building projects that helped researchers understand plankton, study animals on the Serengeti, delve deeper into particle physics, and much, much more.

From trying to understand galaxy evolution to old whaling records and gazelle spotting was quite a journey. In each project, I'd been worried about getting sufficient volunteers, but each time I was elated by the sheer power of people's desire to help.

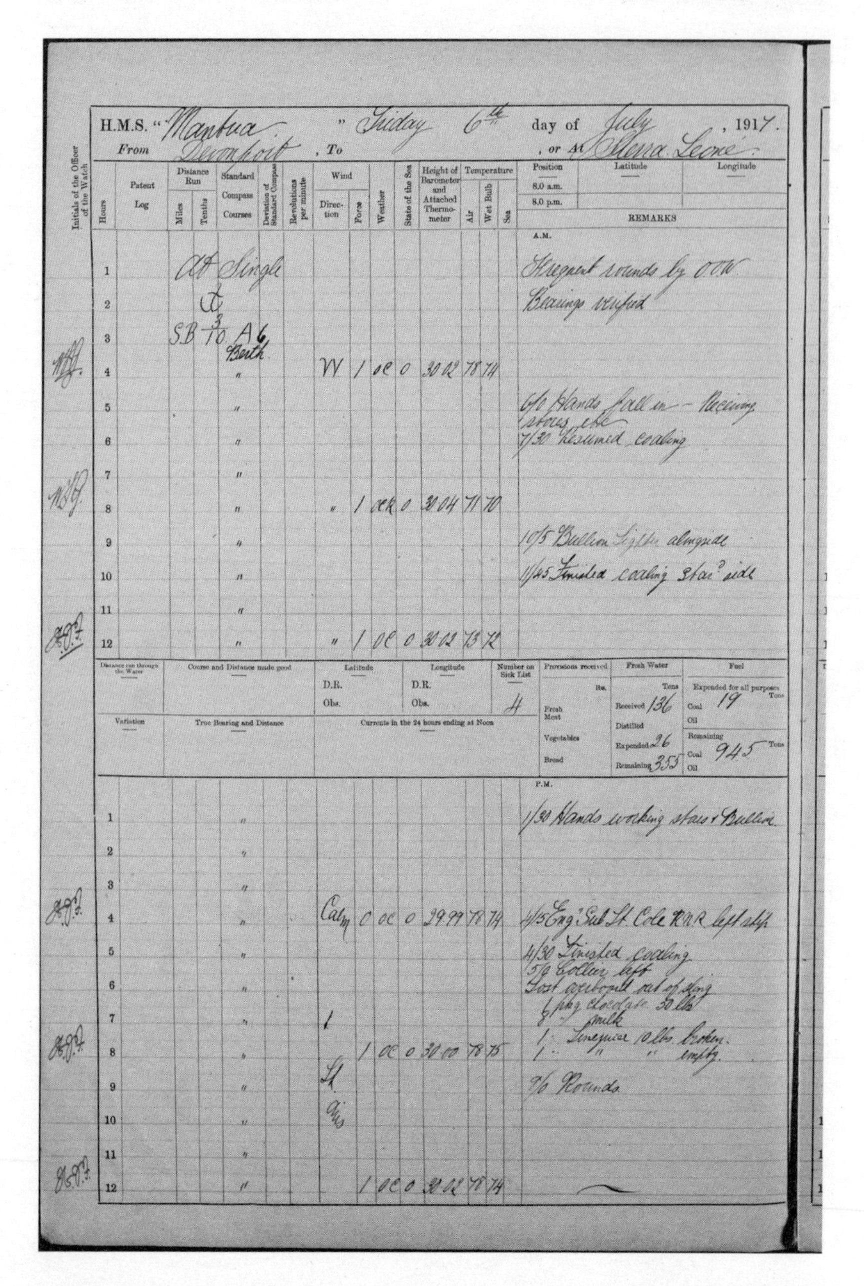

H.M.S. "Mantua" Friday 6th day of July, 1917.
From Devonport, To , or At Sierra Leone

| Hours | Standard Compass Courses | Wind Direction | Force | Weather | State of the Sea | Barometer | Air | Wet Bulb | Remarks |
|---|---|---|---|---|---|---|---|---|---|
| A.M. | | | | | | | | | |
| 1 | At Single | | | | | | | | Frequent rounds by O.O.W. |
| 2 | & | | | | | | | | Bearings verified |
| 3 | S.B 3/10 A6 Berth | | | | | | | | |
| 4 | " | W | 1 | oc | 0 | 30.02 | 78 | 74 | |
| 5 | " | | | | | | | | 6/0 Hands fell in – Receiving stores etc |
| 6 | " | | | | | | | | 7/30 Resumed coaling |
| 7 | " | | | | | | | | |
| 8 | " | " | 1 | ocr | 0 | 30.04 | 71 | 70 | |
| 9 | " | | | | | | | | 10/5 Bullion Lighter alongside |
| 10 | " | | | | | | | | 11/45 Finished coaling Starbd side |
| 11 | " | | | | | | | | |
| 12 | " | " | 1 | oc | 0 | 30.02 | 73 | 72 | |

Number on Sick List: 4
Fresh Water (Tons): Received 136; Expended 26; Remaining 355
Fuel (Tons): Expended for all purposes – Coal 19; Remaining – Coal 945

| Hours | Standard Compass Courses | Wind Direction | Force | Weather | State of the Sea | Barometer | Air | Wet Bulb | Remarks |
|---|---|---|---|---|---|---|---|---|---|
| P.M. | | | | | | | | | |
| 1 | " | | | | | | | | 1/30 Hands working stores & Bullion |
| 2 | " | | | | | | | | |
| 3 | " | | | | | | | | |
| 4 | " | Calm | 0 | oc | 0 | 29.99 | 78 | 74 | 4/15 Engr Sub Lt Cole RNR left ship |
| 5 | " | | | | | | | | 4/30 Finished coaling; 5/0 Collier left |
| 6 | " | | | | | | | | Lost overboard out of sling |
| 7 | " | 1 | | | | | | | 1 pkg chocolate 50 lbs; 1 " milk |
| 8 | " | | 1 | oc | 0 | 30.00 | 78 | 75 | 1 Limejuice 10 lbs. broken.; 1 " " empty. |
| 9 | " | Lt | | | | | | | 9/0 Rounds |
| 10 | " | Airs | | | | | | | |
| 11 | " | | | | | | | | |
| 12 | " | | 1 | oc | 0 | 30.02 | 78 | 74 | |

**Figure 19** Logbook from HMS *Mantua* for 6 July 1917. Weather observations occupy the central columns, with occasional notes on the right, including the sad loss of fifty pounds of chocolate while loading.

The ability to spend a few minutes trying to understand the Universe (or the Earth) was apparently the kind of thing people wanted in their lives, and as we tried new and more complex things I remained in awe of quite how each small effort could add up to something grand.

# 5

# TOO MANY PENGUINS

I don't think I was ever in any serious danger, but as I slid down an icy slope, scraping against freezing rock, my thumping heart would have disagreed. About twenty metres below was the frigid water of the Southern Ocean, its waves turned into slush by icy melt. Scrabbling desperately at the slope through hands hidden within padded gloves, I broke through the snow and hit melt-water running down the rock face. It appeared that I was falling down a waterfall, and the water, I took time to notice, was yellow—liberally infused with a year's worth of penguin guano from colonies higher on the mountain.

It was the penguins that had brought me to this remote place, far from an astronomer's comfort zone. As, somehow, I managed to stop my sliding, I looked up to see my friend Tom Hart inexplicably able to stand on the same slopes that had seen me repeatedly slide towards the ocean. Tom describes himself as a 'penguinologist', and I was supposedly on this trip to assist him in his research.

It's certainly true that Tom looked at home in Antarctica, his uniform of tattered fleece, boots, and rucksack looking much more fit for an expedition to the end of the Earth than to the Victoria pub in Oxford. He is the architect, as well as the caretaker, of an extended network of seventy-five cameras that monitor

**Figure 20** Late summer on the Antarctic Peninsular, with the penguin nesting site now a muddy mess.

behaviour in penguin colonies across the Antarctic Peninsula (the bit of the continent that sticks 'up' towards South America), plus a few scattered across the islands of the vast Southern Ocean. While researchers such as Tom used to visit each colony at best once a year, recording an annual census of the birds, the cameras now take an image every hour (Figure 20).

Thanks to this network, nesting gentoo, Adélie, and chinstrap penguins are under scrutiny they've never faced before. As a result, the complex behaviour of the penguins as they fight for spots to nest, breed, and raise their chicks can be observed and examined. Rather than just having an annual count of each colony, changes in behaviour which might be due to climate change, tourism, or fishing can be recorded, and down here that's important. The Antarctic appeals to Tom, a committed conservationist, because it is protected by the Antarctic Treaty. Though politics interferes, this is the one place on Earth where clear evidence of harm to an ecosystem caused by humans should immediately provoke a response designed to fix the problem. Here, at least, we can make sure we're not ruining the planet.

There are two problems. The first is that while the penguinologists now have access to more data than ever before, they are no more numerous themselves. They face, as we astronomers do, a vast expanse of information encoded in images and have turned to citizen scientists for help. The second problem was the reason we were in Antarctica. There is no readily available Wi-Fi signal that far south, and the cost of transmitting data back to base in Oxford prohibitive. The images are therefore stored locally, and the cameras must be visited to give up their colony's secrets.

Each annual visit also serves as a chance to perform maintenance, to prop up tripods and supports, and to replace the batteries. Tom's life, therefore, is a cross between that of a conventional academic and a travel agent with a penchant for the cold, though I have to admit there's at least a flavour of nineteenth-century-style adventuring somewhere in the mix. The logistical challenge of getting to the cameras is made worse by their placement in obscure and little-visited spots. There is an existing research network in Antarctica and during my visit, members of the team popped in to deliver chocolate to the Argentineans, accidentally stumbled upon the Ecuadorian base, and visited a British Post Office. The scientists attached to these places and the various national programmes have done much to help understand Antarctica, but limiting the survey to penguin colonies easily accessed from these bases would give only a very fragmented picture of what's going on.

Instead, Tom and his team often use what they call 'ships of opportunity' to explore. They are essentially hitchhikers with a purpose, occasionally when needs must and funds allow resorting to hiring a yacht, but mostly travelling on board the cruise ships which take 20,000 or so tourists to the Antarctic Peninsula each year. My trip was on board the Quark Expeditions ship, *Ocean Endeavour*, a voyage on a converted Polish car ferry alongside 180 tourists from around the world who wanted to see penguins and ice. We'd left the southern Argentine port of Ushuaia, one of several places in Tierra del Fuego that claims to be the southernmost city in the world, and survived crossing the Drake Passage, the stormy sea that divides the continents. As I lay as still as possible in a rocking bunk, queasy to the core, I assumed we were being treated to a ship-threatening tempest, but it was nothing so extraordinary; the Drake Passage just is rough.

By the morning of my icy slide, we'd already made it to ten of the network's cameras. Every visit is crucial, because each year brings only a handful of opportunities to get to each site. Miss a camera, and it will fall dark for a year, its batteries drained, leaving a large gap in our knowledge of that site. So the work was important, but it had equally been straightforward. In each place, I'd been able to enjoy spectacular surroundings, in a landscape of towering mountains and looming icebergs under a beautiful late summer Sun.

The sheer diversity of the ice is remarkable, and to me completely unexpected. Most of it glows a deep blue colour in the sunshine, a colour which indicates ice under pressure. Squeezed hard, the air bubbles that normally scatter light and make snow appear white are absent. As we walked, crunching through more recent snow, we left a trail of blue footprints behind us. The white and blue of the ice is broken by dark, almost black volcanic rock formations, and we sailed through passages which divided sheer cliffs. This isn't exactly unexplored territory, but it was surprisingly moving to look from horizon to horizon and see precisely no evidence of human interference.

The aim that morning was to get to the last camera of the season. After this, the oncoming winter would mean that Tom's network would have to survive without him for another year. This one was last for good reason—it is far from the landing spots suitable for a cruise ship full of tourists. I'd reached my precarious position halfway up a cliff via a bouncy ride in a little inflatable boat, a Zodiac, with Tom and one of the ship's expedition staff, Raefe.

Raefe had recently retired from a career with the Australian military, a background which showed up in his meticulous kit, his complete unflappability, and his willingness to be diverted from his day's routine by helping out us scientific hitchhikers. As I clung to the ice, I glanced down to see him sitting with a hand on the tiller of the boat, idling to keep the flakes of ice we'd pushed through from snarling the boat. I was very aware I was making a fool of myself, failing to live up to even the pretence that I belonged in this wilderness, as I awkwardly scrambled up to where Tom was.

He eyed me sceptically, spotting the streaks of guano that now decorated my waterproofs, and announced we were going back to the boat. I'll spare you the details, but getting back was no more fun than making the little progress I'd managed, but all ended well. We found a different landing spot, and Tom set off alone, happy to cope without the presence of an astronomer. I sat and watched him scramble up, and thought about what was going to happen to the data he was bringing back.

The camera on top of the hill hadn't been visited for over a year. For most of that time, images would have been taken every hour, adding up to thousands over the course of the year. During that time, nest sites would have emptied and then disappeared under snow, awaiting the return of the penguins in the spring. Once they returned, then they would have mated, prepared nests, guarded eggs, and raised chicks, all under the snapping gaze of Tom's cameras.

The outlines of this story are familiar to anyone who has ever watched a nature documentary, but it's the details that matter. By flicking through the images from their cameras the penguinologists had noticed something that would have been hard to see otherwise; penguins would occasionally return to their nests even when the colony wasn't occupied. They seem to be keeping an eye on their summer homes, perhaps doing a little light maintenance, but mostly ensuring the spot is still theirs. Nests are defended vigorously when the tightly packed colony is fully occupied, and it seems that at least some of the birds want to get a head start in securing territory.

Those images are strangely moving, a single penguin appearing in a frame or two in the Antarctic twilight. Other newly revealed aspects of penguin behaviour are less romantic. There is a part of visiting the Antarctic that doesn't come across in documentaries, and which won't show up in the holiday snaps of the camera-toting tourists who sailed with us. There isn't a nice way to say this, but penguins absolutely stink. The smell is difficult to describe, being acrid, pungently fishy, and rich and complex like an old brie. Try imagining a heap of manure sprinkled with herring left out in the summer sun and you'll be close.

One of the reasons that the birds smell so bad is their charming habit of carelessly defecating wherever they are. It's common to watch one charmingly tottering along, as photogenic as you like, before stopping to lift its tail and expel a bright white stream of guano with surprising speed and range. A study by researchers Victor Meyer-Rochow and Jozef Gal published in 2003 found the pressures exerted by defecating penguins are up to four times higher than in humans, a fact I mention mostly because they also note that it's not yet known whether the birds are actively choosing which direction to direct the flow.

In other words, it's not clear whether they're aiming, but results from the camera traps suggest there might be a reason for their apparently unhygienic behaviour. When the main colony returns in the spring, some of the nesting sites lie under snow. Until the snow melts, nesting cannot occur, and the main business of the summer can't start. Once it does, the images from the cameras show a hive of activity, but they also show something else.

The areas where the penguins are nesting are often clear of snow long before their surroundings. This is, Tom reckons, not the effect of penguins settling in suntraps, but a consequence of their liberal additions to the local environment. As I'd had the chance to note first hand, snow with penguin guano in it is darker than its surroundings, so it's possible that it absorbs more heat from the Sun—like wearing dark clothes on a hot day. There are other possibilities; it may be that the effect is more direct, and it's the heat of the penguin poop that matters, or perhaps even its saltiness.*

In any case, rather than random bad behaviour it may be that the penguins lack of toilet training is an evolutionary adaption to their environment. To test the idea that it's the darkness of the guano that makes a difference, small plastic discs of varying colour had been left in front of the cameras so the team could see whether they too sank into the snow before the rest melted. This is careful experimental science in the field, enabled by the ability of the camera network to be there for a significant period of time instead of simply making a flying visit.

Interesting things happen when you have a network of camera poles in the Antarctic that happen to have batteries and regular

* This last idea is my favourite, as it means the penguins are essentially salting the icy paths around their homes, just as we do.

visits. Tom is slowly accreting a network of other scientists who are interested in his sites; many of the cameras we visited were newly adorned with small test tubes halfway up the pole. These are pollen traps, collecting the slow drift of material from the air here in a plantless desert precisely because that guarantees an unbiased sample. Colleagues of mine have even talked of converting penguinology stations into detectors for cosmic rays, high-energy particles coming from space.

While running about from camera to camera, and sitting exhausted on the ship thinking only about the next day's efforts, it was easy to close my eyes and imagine myself on a great scientific voyage, exploring the last of the Earth's great wildernesses (Figure 21). Yet people have been travelling to the Antarctic for all sorts of reasons for a long while now; even before I opened my

**Figure 21** A moody day in the Antarctic, with penguins, ice, and our expedition ship in the background.

eyes and reminded myself I was on board a tourist-carrying cruise ship it was obvious from visiting that the Antarctic Peninsula is hardly untouched by human hands. While the still-unforgiving climate, especially the bitter winters, is always going to prevent large-scale immigration, on my trip alone there was plenty of evidence of the use to which human beings have put this great wilderness.

One of the indelible memories I took from the Antarctic adventure is of sailing into a safe harbour known as Deception Island. The island is the rim of an active volcano which pokes above the waves, with its central caldera flooded and a run of flat beaches around the inner rim that provided a site, less than a century ago, for a whaling factory. This whole region was, in the first few decades of the twentieth century, the centre of a trade in seals and whales that butchered now-unspeakable numbers of animals in the Southern Ocean. As a result, the sights of Deception Island include not only rocky peaks and the bluest water you've seen, but also enormous rusting vats that used to contain oil produced by boiling down gigantic carcasses. The vats stand next to rusting bits of dock, designed to service the whaling boats which used to be based here.

The whaling trade didn't survive long. Just a few decades after exploration opened up the Antarctic to fortune-seeking adventurers, business collapsed under the twin pressures of the depression of the 1930s and a fundamentally unstable business; the unfortunate fact is that by then so many animals had been killed that there were few left to profitably hunt. Antarctic exploitation isn't just a problem for the past, though, and with little data available on how we were affecting our surroundings we modern visitors could not afford to be smug. The presence of tourists in Antarctica is both growing and strictly controlled, almost to an absurd degree. Granted access to the bridge as part of my role as a (very) junior penguinologist, I listened in amazement as the crew of the three or four cruise ships in the area negotiated passage and landing sites. Partly they were avoiding landing too many people in any one place—and I should say that everyone I saw was scrupulous in following Antarctic Treaty rules that limited the number of people on shore at one time—but they were also keen to provide the wilderness experience sought by their guests. Our departure from the quiet harbour of Deception Island, for example, was timed so that no one on board saw the next ship slipping in to enjoy an unspoilt experience of an abandoned Antarctic whaling station.

If I sound cynical, I don't mean to. Many of the most moving aspects of the trip for me were the traces of previous human, or at least scientist, occupation. On one stop I was moved to tears by an empty, stone-walled enclosure that didn't even deserve the term 'hut', but which was built by scientists far from home searching for a magnetic pole that turned out to be on the other side of the continent. Tom, meanwhile, was in a spirit of not-quite-irony searching for the remains of Toby the pig, the first pig to visit the Antarctic twice.*

But there was no escaping the fact that one of the things that might be threatening the well-being of the charismatic, if stinking, penguins was our own presence alongside them. Part of the point of our mission was to try and understand human impact

* Toby had sailed south with a Uruguayan expedition, who sold him to a French ship on the way home. Toby's second visit to the Antarctic was thus part of an exhibition organized by the French explorer, Jean-Baptiste Charcot, whose crew were one of the first to deliberately overwinter on the Peninsular. Charcot is remembered now for his superior planning, which allowed for a modicum of comfort which he believed was important for the crew's well-being. There's a photo of him breakfasting—at a carefully laid table—on the ice, butler and champagne on hand, and the ship's crew each morning enjoyed the daily paper from Paris—just distributed precisely a year late.

**Plate 1** The great Orion Nebula, with NGC 1981 (briefly known as Lintott 1) visible at the top of the image, above the smaller patch of nebulosity.

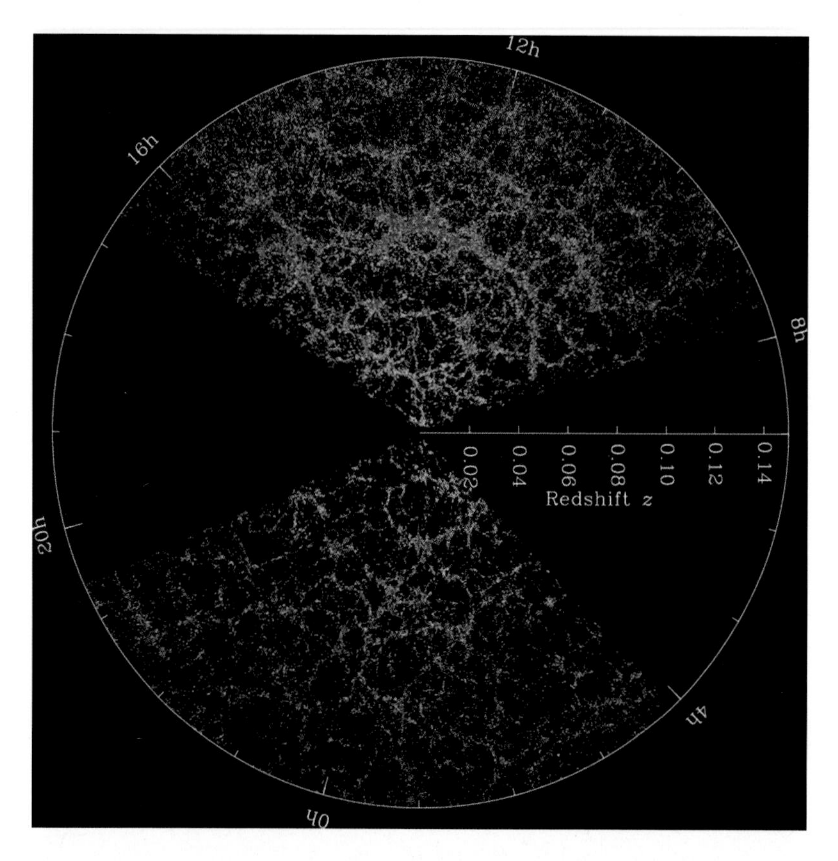


**Plate 2** The Large Scale Structure of the Universe as seen by SDSS. Each dot is a galaxy, whose colour represents the galaxy colour; the Sloan Great Wall is visible in the top segment.

**Plate 3** The protoplanetary disk around the young star HL Tauri, as seen by ALMA. The gaps in the disk may represent disruption of the disk by forming planets.

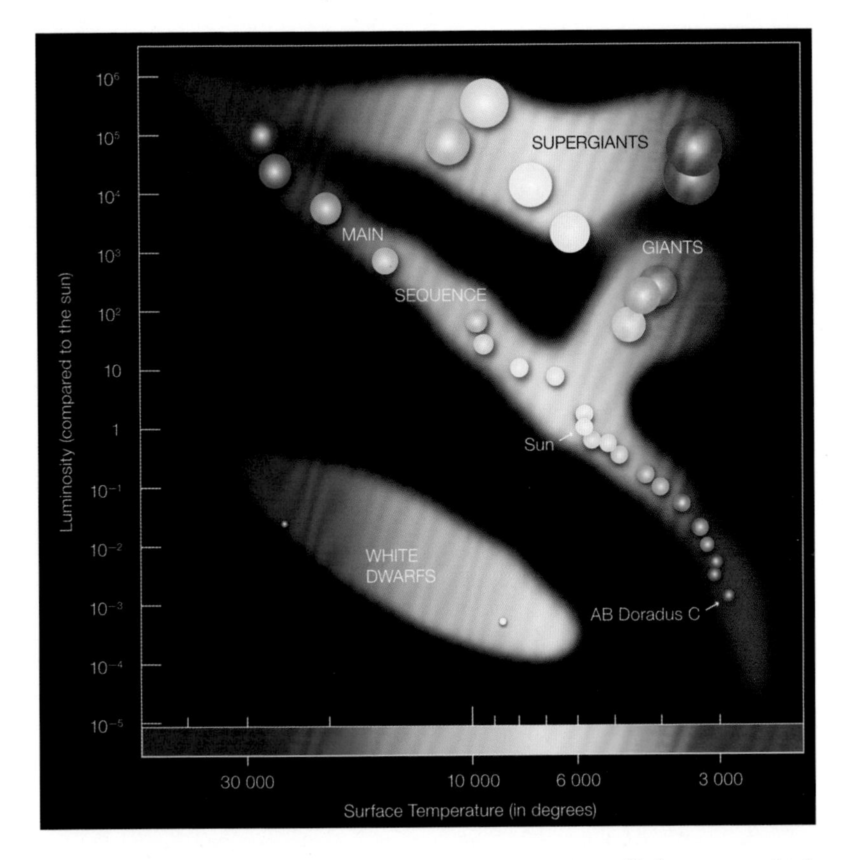


**Plate 4** Schematic version of the Hertzsprung–Russell diagram, which plots brightness against colour (which is equivalent to temperature). Most stars spend most of their lives on the main sequence running from top left to bottom right.

**Plate 5** Two famous galaxies. Top: M51 turns anticlockwise. Bottom: NGC1300 turns clockwise

**Plate 6** 'The Mice' as seen by *Hubble*. This pair of galaxies is in the process of colliding, an encounter which has already produced the long tidal tails visible in the image.

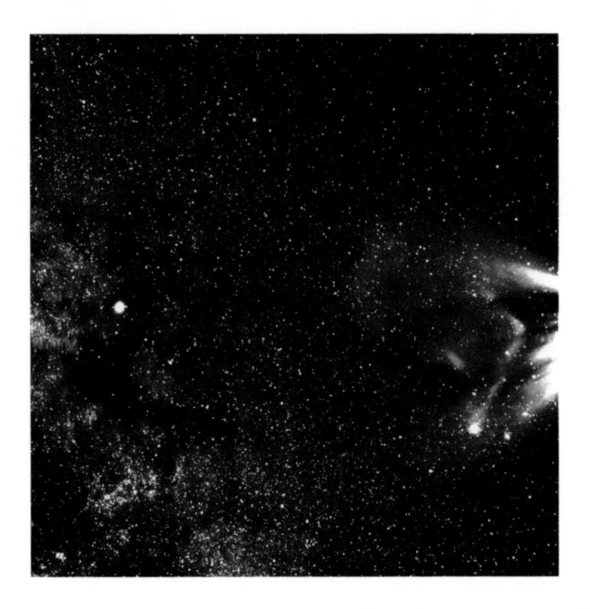

**Plate 7** A coronal mass ejection as seen by the *STEREO* Heliospheric Imager. The Sun is just to the right of this picture, which also shows the Milky Way on the left.

**Plate 8** A magnificent auroral crown as seen from Norway. The green colour is caused by excited oxygen in the upper atmosphere, and the fine structure reveals the complex interaction of the Earth's magnetic field with the solar wind.

**Plate 9** Supernova 2014J in M82, as indicated by the arrow. The red material here is flowing away from the centre of the galaxy, a wind which is perhaps powered by supernovae like this one.

**Plate 10** The 'Red Ring' found by Space Warps volunteers as part of the *Stargazing Live* project. The image was taken as part of a survey by the Canada-France-Hawai'i Telescope.

**Plate 11** Hanny's Voorwerp as seen by the *Hubble Space Telescope*. The complex shape of the Voorwerp—particularly the presence of the apparent 'hole' towards the bottom of the object—is still mysterious.

on the landscape. Tourism was in evidence whenever we were there, for obvious reasons, and science—either by our own efforts or those of the research bases we visited—also had a visible impact. Two hidden threats need to be taken into account, though.

The first is the one you're thinking about. It's hard to mention 'ice', now, without immediately thinking of the melting induced by climate change that is afflicting the world's ice caps and glaciers. The last few summers have been brutal in the polar regions, with the Arctic ice cap frequently so far from its usual extent for the time of year that it's obvious even to most casual observer looking at satellite photographs. The Antarctic Peninsula, too, is warming, although the complexities of the flow of water around here have complicated the picture somewhat. The other threat lies offshore, and comes as something of a shock to those who have come here precisely because it is an unspoilt wilderness.

'Fishing?', they say, as Tom explains his idea that the penguin colonies we are visiting are suffering from the effects of over-zealous human fisheries. The colonies look robust enough, but data from the camera traps show that the numbers of one species in particular is declining. It is the species of penguin that depends most of all on krill, the diverse and nutritious tiny crustaceans that swarm throughout the Southern Ocean.

Krill have been harvested seriously in these waters for the last couple of decades. That sounds surprising, because you've never ordered a krill burger, but in addition to food for fish farms the bountiful harvest produces gallons of sub-standard cod-liver and fish oil. Buy a generic supplement from your pharmacy, and without reading very carefully you'll be competing with the Antarctic penguins for their primary foodstuff.

There is a lot of krill to go round. One source, the Commission for the Conservation of Antarctic Marine Living Resources, points out that the mass of all the Antarctic krill in the world outweighs us humans. Unlike human flesh, though, most of the biomass which exists in the form of krill is eaten each year and then replaced. This rapid turnover makes it even harder to believe that human intervention could make much of a difference, but it actually makes the problem worse.

As the krill disappear, harvested for our consumption, those penguin species which aren't able to adjust to find other foodstuffs suffer. Tom's camera network has already picked up a difference between the resilient gentoo penguins, which have been able to switch from krill to other food, and populations of less versatile chinstraps.

Cameras can only tell you so much, which is why I found myself reaching tentatively into a fridge full of a summer's worth of guano samples. Tom and team have long been in the habit of collecting penguin poop from most of their sites, hoping to marry serious lab analysis with camera data. The cameras tell us how the colony is doing, and the lab work will tell us how at least some individual penguins are doing, bearing information on diet, on health, and on any infections the penguins carry.

First though, we had to get the samples back to Oxford, and as the ship sailed around the peninsula there was a rare opportunity. One of our stops was Port Lockroy, a British base more than a hundred years old and home to the only functioning Post Office on the continent. The volunteer staff from the charity UK Antarctic Heritage, who run the base, act as curators, maintenance staff, wildlife recorders, and more during their stint there, but during the tourist season they also run a thriving gift shop. Must-have items include an Antarctic tartan tie, but what everyone really wants is to send a postcard back to friends at home.

There is, therefore, a working Post Office, though you have to wait for the next ship before mail leaves the base. If Tom and I could package the samples that had been languishing in his fridge, they could be dispatched directly from the base to Oxford. That meant diluting each sample with stabilizing chemical, which meant donning latex gloves and squeezing into the tiny bathroom attached to Tom's cabin.

I've already mentioned the unappetizing smell of the penguin colonies, and the stench of their droppings within the small cabin was unforgettable. It got into one's nostrils instantly, onto clothes, and, I feared, into my skin. After a while on the production line, handling tubes Tom passed to me while pointing them as far from my nose as possible, I pleaded for a break and headed out to get coffee.

I got back to the cabin to notice one of the ship's efficient crew fiddling with pipes in the corridor outside. I was just explaining this to Tom when a knock on the door revealed the ship's purser, in pursuit of an unearthly smell that was disturbing those paying passengers who were trying to have the holiday of a lifetime around our scientific expedition. Apparently he wanted to see Tom's bathroom, where he feared the smell now working its way through the ship's air conditioning originated.

I don't really know how to describe what we looked like, two unshaven researchers with blue hospital gloves, inane grins, and the realization of what we'd done slowly showing up in our expressions. Somehow, between our shock and his confusion, we agreed he should come back and inspect the plumbing later and got on with the job. I've never worked faster in my life, and somehow we got the samples safely packaged before a more forcible intervention arrived.

Reflecting on the morning's events in the ship's bar later, wondering if people were avoiding me because of a lingering stench, I realized just how close to astrophysics Tom's research was. Obviously, I'm rarely called in to deal with galaxy excreta, at least directly, but Tom (with my inept help) had become, while still doing science, a professional and an expert in data collection. His command of the moving network of ships and people, and the resulting spread of cameras and data, reminded me of the unsung heroes of the Sloan Digital Sky Survey, the engineers and astronomers who spend huge amounts of time gathering the data on which the rest of us depend.

The fact that we'd rushed to treat the samples before the smell contaminated a cruise ship won't ever be mentioned in a scientific paper. The careful day-to-day diplomacy that ensured that a ship employed on quite some other purpose delivered the team to each of their cameras is reflected in an unbroken data series, but won't ever be commented on in formal publication. And the fact that Tom was able to leap up that hill, and I wasn't, won't be recorded anywhere but in the pages of this book.

Similarly, without people who understand how to make the telescope perform, to keep the camera operating at peak performance, all the science that uses this data is measurably poorer. Without people who are really good at being Sloan's 'cold observer'* my measurements of galaxy properties would be less accurate. It's rare, in science, to pay much attention to these hidden parts of the process, which since the nineteenth century have become increasingly professionalized and, for most of us as we've entered the digital age, increasingly remote from our day-to-day lives.

When you go to that much effort—whether in the surprisingly chilly New Mexico night or the much more predictable cold of the Antarctic—it's important to make the best of all of the data you can obtain. For Tom and his team that means sharing the images his camera network takes, and which they go to such

* This is a real job title; the cold observer is out with the telescope while the 'warm observer' is inside with the electronics.

lengths to return to Oxford, with the entire world, via a Zooniverse project called Penguin Watch.

Penguin Watch is perhaps the simplest of all our projects, so straightforward that I know 5-year-old children have taken part. Presented with an image from the cameras, all you have to do is count and then click on the penguins. This information seems almost banal in the context of a single image, but over time we learn how colonies shrink and expand, how much time individual penguins are spending out at sea, and even what their feeding behaviour is like. These details can then be compared to weather and climate records, to changes in fishing permits and protected areas, and to visitor numbers to get a sense of what's really happening.

For that to work, the data must be accurate. An individual penguin counter can make a mistake, and so the key is to combine everyone's penguin counts to produce a consensus. All of our citizen science projects depend on multiple people looking at the same data, but there are a few maddening quirks about penguins that make it more difficult. First—and you might have to take my word for it—there are a lot of them. We, of course, are after an accurate, scientific count, but some of the cameras capture hundreds of penguins in a single image. That's not ideal, obviously, but the cameras are set up once and then left for the year and so when things change, so does their view.

The problem with having too many penguins is that people baulk at counting them. Being presented with an image of hundreds of the critters is, for many people, more annoying than interesting. Our designers and developers know this, and so Penguin Watch reassures you that after your penguin count reaches thirty it's OK to move on. (It turns out there are people out there, many of them attracted to Penguin Watch, who deeply resent this message. They are people who like order, who like completing a challenge no matter how many penguins it involves. And so we learn once again that people are complicated.) For these busy images, therefore, few people cover every penguin. Some click the front row, others a cluster near the back, and so on and so forth.

What we're left with once many people have seen each image is a mosaic of things that at least one person thought was a penguin. What the team need is a list of likely penguins, ideally with some sense of how likely each possibility is to be real. This requires some careful data handling, but the basic idea is simple—if two people mark a penguin in roughly the same place, then each marking counts as a 'vote' that there's a penguin there. The nice thing about this is that the researcher working on the data, has only got to make two decisions. The first concerns how close together two markings have to be for them to be in 'roughly' the same place. That can be found by trial and error. If you have a few images that experts have gone through, then you can just adjust the parameter until you get results that look pretty good. If you fail, then you need more people to look at each image so that you get more data. This is essentially what we do when testing a project.

That's the easy part. If you want to make it complicated, there are reams of computer science papers that deal with this sort of clustering problem, and plenty of researchers who will make it more complex for you. The degree of proximity required to have the algorithm decide that two marks refer to the same penguin need not be a constant, for example, but could depend on how far the markings are from the camera, or the time of day, or how many other penguins are in the image, or a host of other variables.

In general, though, because Penguin Watch volunteers are pretty good, there's not much need for a complex solution to the problem. The second decision you have to make is much more difficult. How many people have to have marked the same spot for us to conclude there is a penguin there?

We want accurate data, so the temptation is to say that we need lots of people for each penguin. That'll produce a set of places where we're really, really sure a penguin is—but we'll miss most of them. If we want complete data—if we want to catch every penguin going, even the ones at the back disguising themselves effectively as rocks—then we need to relax, make the algorithm less picky, and include places where only a few people say there's a penguin in the final list.

That will decrease the accuracy, so there's a trade-off to be made between accuracy and completeness; in a problem like this, you have to choose which you care more about. This turns out to be a general feature of this type of problem, and for different scientific problems you might pick different combinations. If you wanted a few excellent images of penguins, for example, you might go for high accuracy and low completeness. If your research called for an upper limit on the number of surviving penguins, then completeness becomes more important than accuracy.

But this isn't the end of the story. We haven't used all of the information we have to hand, and in trying to squeeze more from the data, in order to do justice to all the hard work that went into collecting it, things get interesting. So far, we've treated everyone's classifications as being of equal value, but it must be true that some will be better, or more diligent, at penguin counting than others. For example, we know a large proportion of people who take part in Penguin Watch in particular have 'Mom', 'Mum', or 'Dad' in their usernames; it seems reasonable to assume that these represent households where participating in science by counting penguins is a family activity, involving children too young to have their own account.

It's possible too that these kids are less good than their older counterparts at the task, and thus we should pay less attention to them. On the other hand, I could easily believe that the combination of growing up as digital natives, smaller hands, and the insatiable desire for repetition that characterizes most 5 year olds of my acquaintance might make them killer classifiers. We don't have to decide in advance; with only a small number of images labelled by experts, we can find the people who are best at penguin counting, and pay attention to them.

It's a simple and obvious idea. The ability of websites to understand us based solely on our interactions with them is one of the things that drives the digital world. If Google can discover just by watching the information I happen to give it where I work (after a brief, embarrassing period during which it insisted the Lamb and Flag pub was my office), and if Facebook can serve up the memory of the long-forgotten party I want to see each morning, then surely discovering whether I'm any good at penguin counting by asking me to count penguins is the digital equivalent of child's play.

Once we start thinking like this, there's much we can do to improve the efficiency of citizen science projects. As well as paying different levels of attention to individuals based on their performance, we can start thinking about how to combine human and machine classification, or even how to manipulate what people see to make them more likely to stay on the site. As you might imagine, this gets tricky quickly, and to think about the possibilities we need to return to space, and to looking up at the night sky.

# 6

# FROM SUPERNOVAE TO ZORILLAS

One of the things I like about Oxford is that the sky is still pretty dark. I don't live far from the centre, and a relatively benign street-lighting policy means that as I wheel my bike into the garden after a late night in the office (or the Lamb and Flag) I can look up and see the stars. Partly, I like being able to mark the passing of time in the changing display of constellations, but I'm looking too for a glimpse of unchanging infinity. It's nice to be assured that whatever crises are happening here on Earth, the vastness of the Universe is there, beautiful and silent and unchanging. It's clearly not just me that feels like this; from Immanuel Kant, filled with 'admiration and awe' by the starry heavens above him, through Walt Whitman's protagonist who braves the 'mystical moist night air' to look up 'in perfect silence at the stars', plenty have gazed on the sky for a bracing dose of cosmic perspective.

The effect isn't limited to looking at the sky directly, either. A good 8 per cent of respondents to the first survey we carried out of Galaxy Zoo volunteers said they were participating in the project because they enjoyed the opportunity to 'think about the vastness of the Universe'. The bad news is that modern astrophysics is hell bent on making it clear that the unchanging, ever-enduring sky is anything but that. It changes, and increasingly astronomers are paying attention to this changing sky, an attitude which is bringing exciting discoveries. Whereas in the twentieth century hacking at the frontier of observational astronomy meant stretching to new wavelengths, to radio astronomy, and to high energies, in the twenty-first it means paying attention to how things change with time. This would have surprised our predecessors. Of course, a few things have always been known to change—the phases of the Moon, the positions of what the ancient Greeks called 'wandering stars' which we now call planets—but at least regular and increasingly predictable patterns could be discerned. The idea of sudden, violent change would have been deeply shocking; it seems to have been at least one reason that comets, appearing without warning and then vanishing again, were seen as omens.

The shock must have been worse when apparently simple, straightforward, and familiar objects misbehaved. Comets belong to the Solar System, and whizz past with their message of doom, but supernovae are another thing altogether. They appear as new stars, shining brightly for a period of a few weeks or months before fading forever. In May 1006, for example, the otherwise obscure constellation of Lepus—the hare at Orion's feet—suddenly boasted a bright new jewel. Bright enough to be easily visible in daylight, it was recorded by observers in Europe, China, and Japan.

For a few short months it was the brightest thing in the sky other than the Sun and Moon, exceeding even the most brilliant apparition of Venus. Similar events were noted in 1054, in 1572, and in 1604. This last event was observed by Kepler, who reached for an explanation in terms familiar to anyone at the time, pointing to the event as a possible analogue of the star that the Bible tells us shone over Bethlehem.

That 1604 supernova is still the most recent to have been observed within the Milky Way; centuries of instrumental advance since the invention of the telescope have been denied the chance to study such an event close up. The closest we've come is a 1987 event in the Large Magellanic Cloud, the largest of the Milky Way's satellite galaxies, which was brilliant enough to be seen with the naked eye and was observed with every instrument available.

Supernovae are, though, powerful enough to be seen in distant galaxies. Most such events mark not the birth of a new star, but the death of an old one much more massive than the Sun. Most stars spend most of their life converting hydrogen to helium, before moving through stages where they burn heavier and heavier elements, all the way up to iron for the most massive stars. When such a star runs out of fuel at its core, or when the temperature is insufficient to start the next set of reactions, the nuclear fusion that has sustained it ceases. This is a problem as the light produced in normal nuclear reactions will stream outwards, encountering atoms in the stellar atmosphere and producing a pressure which supports the star's outer layers.

Once this radiation pressure which normally prevents gravity from taking its course vanishes, the star collapses in on itself. This dramatic event can liberate enormous energies; a typical supernova shines with a hundred billion, billion, billion, billon watts. I don't have a good comparison for this, but it's enough to match the power produced by all of Earth's power stations for at least a couple of billion years, or by more than a billion billion suns; at its peak a single supernova will easily outshine the collective might of an entire galaxy's stars.

It's this brightness that means that supernovae can be seen from distant galaxies, and that makes them extremely useful.

One particular kind of supernova, which astronomers call 'type 1a' supernovae, can be used to measure the expansion of the Universe itself. These rare events have a slightly more complicated story than most, being the product of two stars rather than one. The most likely scenario for the formation of a type 1a supernova starts with a binary system consisting of a pair of Sun-like stars.

Normally, stars the size of the Sun don't go supernova. When hydrogen is exhausted at their cores, they can begin burning helium. When this happens, the equilibrium between gravity and the pressure pushing outwards is disturbed, and the star will expand to become a red giant. The core's helium too will eventually be exhausted, and the star will shed its outer layers, producing at the end of its life nothing more than a transient, if beautiful, planetary nebula from the gas lost in these outer layers and the dense, cooling remnant of the core known as a white dwarf. Left to its own devices, such a relic would cool slowly; nuclear reactions will have ceased.

In a binary system, a more interesting future is possible. The more massive of the two stars will undergo evolution as normal, ending up as a white dwarf. If the second star enters a red giant phase and is sufficiently close to the white dwarf, material may be pulled by gravity from the still-shining red star onto the surface of its dead companion.

At first, not much happens, but as the mass of material accumulating on the surface of the white dwarf increases there will come a point where it will reignite, and the star will, briefly, shine brightly once more. This isn't a slow and stable process, but rather a chain reaction in which all the accreted material rapidly becomes involved. From almost nothing, the star will shine more brilliantly than it ever has before, appearing just for a short time as a spectacular supernova.

What's more, the timing of this runaway nuclear reaction will mostly depend on how much fuel has built up; there will be a common mass required to ignite new reactions whenever and wherever in the Universe this happens. That means that whenever a type 1a supernova explodes, it will do so with the same brightness, shining out into the Universe with a standard luminosity.

In the few sentences above, I've ridden roughshod over about twenty years' worth of work of many of my colleagues. While type 1a supernovae are always roughly the same luminosity, exactly how bright any particular supernova will be depends on many different factors, including the composition of the material involved. Luckily, by looking at the details of how the supernova brightens and fades, we can adjust (we think!) for most of these effects.

Astronomers like such objects. If you know how bright something really is (how luminous it is), and can measure how bright it appears, then you have a measure of distance. If I told you that I was holding a lamp containing a sixty-watt light bulb, you'd be able to hazard a sensible guess as to whether it was a metre or a mile away from you. Such 'standard candles' are the main tools by which we measure distances; the Cepheids discussed in Chapter 2 are a famous example. Type 1a supernovae are merely the latest in a long line of such objects, but they are especially valuable because of their great luminosity. Once astronomers had captured enough of them it turned out they had a surprise for us.

Two big research efforts in the 1990s set out to systematically discover and then follow up these supernovae for cosmological purposes. One was led by Saul Perlmutter in Berkeley, and the other by Adam Riess at Johns Hopkins and Brian Schmidt at the Australian National University, though like much of today's astrophysics there were large teams at work in both cases. The idea was to measure enough distant type 1a supernovae to get a sense of how the expansion of the Universe was changing.

Hubble's constant—the speed of expansion of the Universe—isn't really a constant. The speed of expansion has been changing since the beginning, affected, for starters, by gravity. The pull of the matter in the Universe acts as a drag on the universal expansion, and so both teams of supernova hunters expected to be measuring a slowing down of this expansion.* Instead, they found a remarkable, deceptively simple result. To their great puzzlement and surprise, both groups found that many of the type 1a supernovae were fainter than expected. Stranger still, the effect increased with distance; the further away a supernova was, the fainter it was compared to the expected brightness.

This sounds like it should be some systematic effect, rather than anything real. For a while, plenty of people believed that the results could be explained by light being affected by its passage towards us, perhaps absorbed by dust, but observations of supernovae at different wavelengths ruled that out. Remember, too, that when we look deeper into the Universe, we're looking back in time, covering billions of years of history. Maybe type 1a supernovae were different back then. Certainly, the stars which create them will have been different from those around us today, having formed from pristine material produced in the first few minutes after the Big Bang which is almost entirely hydrogen and

* The technical details don't matter too much here, but the secret is to measure the distance to the galaxy hosting the supernova in two ways. Measuring the apparent brightness of the supernova is one way, but we can also measure the redshift in the light from the galaxy caused by the expansion of the Universe while it was en route. Combining the two for supernovae at different distances gives a measure of how the expansion rate is changing.

helium; the heavier elements are all produced in stars and mixed back into their surroundings upon the death of the star. Early in the Universe's history there just hadn't been much time for stars to live out their lives and then die, and therefore pollutants such as carbon and oxygen are relatively rare.

We (more or less!) understand stars, though, and how this change in composition would affect them. No matter how people tried, it didn't seem like astronomers could get away with blaming the unexpected results on stellar properties. Lots of careful work instead led cosmologists to a more radical conclusion. The supernovae, it seems, are fainter than expected because they really are further away than expected. The expansion of the Universe has sped up during the time that light has been travelling from these distant events to Earth. Instead of the universal expansion slowing down under the influence of gravity, it is speeding up.

To say this is confounding is to understate the case. We need some sort of 'anti-gravity' capable of acting on the largest of scales and speeding up the expansion. Worse, the measurements from large surveys of type 1a supernovae, alongside many other strands of evidence suggest that this anti-gravity is utterly dominant, accounting for about 70 per cent of the energy (more technically, the energy density) in the Universe today. This accelerating force has come to be known as 'dark energy', presumably because we need a name as confusing as the thing itself (I don't know what it means for energy to be 'dark'), and understanding it is, in my opinion, the largest outstanding problem in physics.

I admit that the quest to reveal the nature of dark energy is a long way from everyday concerns, but the entire fate of the Universe depends on it. Its influence increases as the Universe gets larger, which means that because of dark energy the acceleration we observe today will continue. Rather than collapsing back in on itself towards a Big Crunch,* or slowing its expansion as it approaches some maximum size, the Universe will continue expanding forever.

While this seems cheering at first, what we face on a cosmic scale is a long, sober, and increasingly dull retirement. More stars are now dying than are being born each year, and as the expansion continues we will eventually reach the moment when the last star is born. All that remains after that is the long, slow dying of the light as first the most massive and then the smaller stars die one by one, their lights extinguished. Our Universe's far future contains nothing more than an ever-expanding sea of radiation.

Worse, from an astronomer's (admittedly long-term) point of view, is the fact that the acceleration due to dark energy pushes objects over the cosmic horizon; the proportion of the Universe which we can see is decreasing due to the increased rate of expansion.† So what is going on?

We don't know. Until the discovery of dark energy, physicists had been content with four fundamental forces (gravity, electromagnetism, and the strong and weak nuclear forces), but none can explain accelerated expansion. A natural candidate for a repulsive force emerges from some of the ways that quantum physics describes the behaviour of empty space, but this 'vacuum energy' has to either cancel out or else have a minimum strength around a thousand billion, billion, billion, billion, billion, billion, billion, billion, billion, billion, billion, billion, billion times stronger than we observe. Unless something's wrong with the basics of quantum theory, we can rule this out as a source of dark energy by observing that the Universe has yet to tear itself apart.

* I've always liked Douglas Adams' suggestion that such an event should be called a Gnab Gib, as it's a Big Bang but backwards.

† This is one reason why it's important to invest in astronomical research right now.

Faced with celebrating this, the largest error ever achieved by science, my theoretical colleagues have not been idle. There is no broadly accepted theory (it would be cruel but not that inaccurate to say there are few detailed theories that have more than one or two adherents), but there are plenty of ideas. The flourishing of creative ideas in response to a mystery is heartening, and browsing the research journals you can take your pick from ideas that are relatively mainstream—some tweak of the vacuum energy, a scalar field left over from the start of the Universe—to the speculative. The latter possibilities are maybe the most exciting, often involving a direct challenge to the foundations of Einstein's general relativity and its ideas about how gravity works. (Not a game that traditionally ends well for the challenger, but I suppose there's always some hope that this time we'll catch Albert out.)

To concentrate, as a thousand articles and not a few books have done, on the intellectual games of theorists, no matter how diverting or elegant they are, is to miss the point.

One of these ideas will win out, but not because of a sudden theoretical breakthrough. There will be no dropped chalk at the end of a lecture to a stunned audience, and no one will be leaping out of a bath shouting 'Eureka!' What's needed, desperately, is more data. If we knew, for example, that the strength of dark energy was changing over time, that would rule out many of the available theories and give researchers something to aim for. What's more, the observational route to this is clear. What we need are more supernovae. The discovery of more distant examples would mean that we could compare the past effect of dark energy with present-day values, and the discovery of more nearby explosions means that we can take better account of systematic effects.

This search has recently become even more important. Just as it looked like the results from many different cosmological probes were converging on a single solution to the parameters that control the Universe's evolution, an intriguing set of results suggest we might not be done yet. Measures of Hubble's constant- the rate of expansion of the present day Universe made with supernovae are giving a higher answer to those derived using methods which depend on measurements of the cosmic microwave background, which suggest the Universe is expanding more slowly. In a well-behaved Universe, the two should agree with each other, so this is puzzling.

It could be that there's nothing to worry about. The difference is small enough that the 'tension', as it's coyly termed, could just be due to chance, similar to flipping a coin and getting three heads in a row. In such circumstances, it's probably premature to conclude that the coin is biased towards heads. In the case of these separate measurements, there seems to be a little more than a 1 in 100,000 chance of the difference being a coincidence; not enough for scientists to be sure that's it's real, but certainly enough to worry about. Whole conferences have now been dedicated to the problem, and both groups—those that study supernovae and those who stare at the cosmic microwave background—are adamant that there's no simple explanation. Either we don't understand the early Universe properly, or something is seriously wrong with our cosmological models, or supernovae are odder than we think.

All of those possibilities are exciting, and in each case we need more data, and so both understanding this intriguing result and getting a critical clue to the nature of dark energy—the key problem in twenty-first-century physics— depends on our ability to find changes in the sky. We've already seen that finding planets—and maybe (though probably not) aliens—is essentially a problem of watching things change. So is keeping the Earth safe from killer asteroids, or detecting the relics of the earliest days of the

Solar System that lurk beyond Pluto. Our understanding of stars (and whether they can support planets with intelligent life) depends on understanding their variability; the Sun has sunspots, and some other stars at least have starspots. And it's not too much to hope that one day soon we might watch the centres of nearby galaxies flicker as material falls into their central black holes.

Given this long to-do list, it's not surprising that telescopes all over the world are being converted to look for new transients. There is hardly a modern survey that doesn't have looking for changes in the sky as at least one of its goals, but to do the job properly dedicated facilities are needed. Optical systems need to be stable to allow images taken days, months, or even years apart to be compared, and instruments on the look-out for changes need to have as wide a field of view as possible. In the most extreme cases, cameras and computers might need to work fast to trigger alerts so that other telescopes can follow up on discoveries.

To get an idea of what a modern transient-hunting machine might look like, you could travel to a hitherto obscure peak in the Atacama Desert in northern Chile. The desert has long been recognized as one of the best places on Earth for observatories, high above the often cloudy coast but lower than the snowy peaks of the Andes. Look at a satellite photo of the area, and the most typical sight is a clear strip between two belts of cloud, and it's here that many of the world's largest telescopes are placed. On a mountain called Cerro Pachón, construction of a new eye on the sky is underway.

This is the Large Synoptic Survey Telescope, or LSST, whose mirror I encountered earlier in the University of Arizona's surreal football stadium-based mirror lab. Staring at a distorted version of myself in the newly shiny surface, it was hard to imagine that it would ever get anywhere near being ready to ship data to the world's astronomers, but now first light—the moment when the first pictures of the sky are taken—is just around the corner. The initial images with a temporary, commissioning camera are due in 2021, and the survey proper will start, all being well, in 2023.

It's hard for me not to be slightly scared by the prospect. The roots of the LSST project go back almost two decades, when the first plans for such a telescope were hatched. Even then, it was clear that for all the clever optics, the biggest challenge would be dealing with the data such a survey would produce. After full operations start, LSST should produce about thirty terabytes of images a night, more each night than the *Hubble Space Telescope* produced in its first fifteen years. That's just the static images, though, and it's the numbers of expected transients which are which are truly frightening.

Nothing like LSST has ever been built before, so predictions are uncertain, but subscribing to a service that provided a text message every time LSST detects a change would leave you waking up to at the very least a million text messages every clear night. Most would be routine changes—viewed with a telescope as large as LSST, a very large number of stars will vary in brightness—but hidden in the stream will be everything you can imagine. If type 1a supernovae are your thing, there will be plenty hidden in the data if you can only find them.

One solution is to depend on machine learning. Scientists all over the world are preparing 'brokers', little software helpers which will listen to the great stream of data flowing from the observatory and shout loudly when they spot something interesting. For the most interesting or useful transients, I suspect we'll see competing brokers from different teams, announcing the highlights from LSST's transients, either loudly to their world or more quietly to their creators. Like an electronic version of an old trading hall, the advantage will accrue to those who can best filter information or make sense of the cacophony.

Not all those acting as brokers will be machines. Where there is a wealth of data to be organized and sorted, the experiences recounted earlier in this book have taught me that there might be a place for citizen science; the Zooniverse hosted its first supernova-hunting project back in 2010. The data rate, provided by a reconditioned telescope on Palomar Mountain in California now pressed into service as the Palomar Transient Factory, was a little more tractable, but the principle was the same. The telescope scanned the sky, and a computer checked each night's images against a set of standard images. The few thousand such candidates a night were uploaded to our website, and a dedicated band of a few thousand volunteers jumped onto a dedicated website each day to sort through them.

They were fast, collectively analysing a night's worth of data in just fifteen minutes, and they were accurate. We were able to broadcast their classifications to observers stationed around the world, and add newly confirmed supernovae to the cosmological harvest. As the survey progressed, though, these classified supernovae also provided new training data for use by would-be supernova-hunting robots. Eventually, an extremely bright student in Berkeley, California produced a trained machine-learning solution that performed accurately enough to satisfy the astronomers running the survey, and as they preferred clinical algorithmic precision to messy and confounding citizen science our project was no more.

I'll return to this project later, as I think the experience of the volunteers who took part has much to tell us about the future of citizen science in general. For now, though, let's continue to think like transient-hunting scientists, and worry about getting hold of as much data as possible. Proof that relying solely on their machine was a mistake arrived at Earth on 21 January 2014, and was first announced by an unusual team from a truly unlikely place.

London is a terrible spot to put an observatory. If you had to pick a site among the glitz and glare of the brightly lit metropolis, the very worst place would be in the centre of the West End. The second worst, though, would be along one of the capital's main roads—alongside the A1 as it cuts through built-up North London, for example. Yet if you drive north on the A1 and look left just at the right time somewhere in Edgware, you'll spot the gleaming domes of the University of London's Mill Hill observatory.

It's a long way from a pristine Chilean mountain top, but that's OK. The observatory exists primarily as a teaching tool, giving students on astrophysics courses at University College London experience in carrying out astronomical observation and data reduction. While the largest telescope is still the beautiful Radcliffe refractor, now more than a century old, it's the modern telescopes clustered around it that get the most use.

Back on that fateful January night, Steve Fossey—doyen of the observatory's teaching labs since well before I was a PhD student at University College London—was scheduled to give a practical introduction to the telescopes to a bunch of undergraduates. Light pollution isn't the only problem with the site, though, and clouds closed in overhead as the session was getting going. As the students took a break with pizza, Steve slewed one of the smaller telescopes over to one of the last clear patches, a region in Ursa Major that contains the nearby galaxy M82.

M82 is known as the cigar galaxy—it is a spiral viewed almost edge on, presenting itself as a thin needle of light on the sky. As that night's image appeared on the screen, Steve noticed a new, bright star located at one end of the disc, something that definitely wasn't in archived images of the same galaxy. From eating takeaway pizza, four students—Ben Cooke, Guy Pollack, Tom Wright, and Matt Wilde—were suddenly following up on what proved to be the supernova discovery of the twenty-first century so far. The clouds were closing in, and the students and Steve rushed to get confirmation images using filters that exposed the camera to different colours. Clinching evidence came when they used a second telescope on the site to take an image of the same galaxy, and saw that the supernova was still there. It wasn't an instrumental error, or something weird happening in the camera; it was real (Plate 9).

From there things moved fast. The standard procedure is to report such a discovery to the wonderfully named Central Bureau for Astronomical Telegrams in the US, who announced the discovery to the world. Within hours of the initial discovery, telescopes around the world had observed what the University College London team had found, confirming that supernova 2014J (as it was now known) was not only real but a type 1a. The opportunity to study this most important type of explosion up close—or at least at a distance of only eleven and a half million light years—was unprecedented, and it became one of the most observed objects of the twenty-first century.

The discovery of such an object by pizza-munching students is a great story, and it was wonderful to see Steve's sharp eyes get some recognition, but the truth is that they should never have had a chance. Automated surveys had caught the supernova before it was observed in London, but the routines used to scan for interesting transients didn't catch it and so didn't sound the alarm. This seems odd. The supernova is incredibly obvious in images—it's the bright star that wasn't there in 2013—but this is only true for human observers. My guess is that the training sets used to send machines hunting for transients didn't include anything this bright, and so the computers had 'learned' that anything that obvious couldn't possibly be real. And so the supernova remained unfound.

There are, to my surprise, at least ten images of the M82 supernova from before the discovery, including some from amateur astrophotographers who either didn't process their data straight away or who didn't know the galaxy well enough to recognize that the star was new. Mostly it was the former; even amateur astronomers with (advanced) backyard telescopes now leave looking at the images to the daytime rather than viewing them as they come in. Such images are, after the fact, still incredibly useful; it turned out that the rise to peak brightness was more rapid for this event than normal, indicating some unexpected process at play, a result that is still causing debate among experts.

One of the surveys that imaged M82 during the period when the supernova was visible but not yet known was the Palomar Transient Factory which fed data to the Zooniverse's supernova project. Had our supernova project still been operating I'm sure we would have caught it, and quickly. There is, of course, no real impediment to adapting the machine-learning routines used to include objects like this one; if I'm right that it was the brightness that made it difficult, one could simply train on as many bright supernovae as necessary. (If there are enough examples in both the Universe and our survey of it, that is. I'll talk about truly rare objects a little later.) The point, though, isn't that one couldn't possibly have designed a system which wouldn't have failed in this way, it's that no one did. When dealing with a complex problem, and real, messy, noisy data, anticipating every possible eventuality is difficult. Ensuring that every case is covered, every loophole closed, and every unusual object anticipated is impossible. Preparing a training set that reflects reality is next to impossible.

We could continue to bet on improvement in machine learning. We might have missed this supernova, for example, but there will be others. Certainly there's plenty of research funding going into making such systems work better. I prefer, though, to acknowledge the limits of any system we build and look to combine the best of automated scanning, which brings speed and consistency, and the quirky responses of adaptable citizen scientists, capable of going beyond their training.

To understand how this might work in practice, we've recently revived supernova hunting as a sport at the Zooniverse. This time the data comes from Pan-STARRS, a camera and telescope which sits on top of Mauna Kea and which was built to hunt for asteroids. It does a pretty good job of looking for supernovae along the way, and once a week we release a week's worth of data to a growing community hungry for discovery. The set-up is even simpler than before: after reviewing a few example images we simply ask volunteers whether a new discovery looks like a supernova.

This time, though, there's a machine running in parallel. It was built by Darryl Wright. Darryl's now part of the Zooniverse team at the University of Minnesota, but when he was a PhD student working with the Pan-STARRS team at Queen's University in Belfast he was asked to review candidates by eye himself. Instead, he took an online course in machine learning and ended up training a neural network to classify the things instead. With the new project, we could compare Darryl's machine's performance with that of the volunteers, and work out which was best.

Once we agreed what 'best' was, that is. As in the penguin-counting example, it's a nebulous concept, and how one might use it probably depends on what kind of science you're trying to do. If you want to make a detailed study of only a few supernovae with the largest of telescopes, then who cares if you miss most of them—all you should watch for is the accuracy of those that you do capture. An inaccurate classification will cost you valuable observing time and earn you the wrath of other astronomers who want the telescope for themselves. On the other hand, if you're trying to understand the properties of a population of objects, then you might not care if one or two false alarms sneak through, and would accept lower accuracy in exchange for catching more of the supernovae in your net.

This is a common trade-off in this sort of classification problem, but it turned out not to matter too much. We quickly found that for almost any realistic case, combining human and machine classifications outperformed any result provided by each alone. Working alongside our robot friends makes us more productive, but input from humans also helps them get better at classifying.

The really great thing about this result is that there's nothing especially clever about it. The citizen science project asks a simple question to a small group of volunteers, and we're not doing any sophisticated data analysis, just believing that the majority of people who answer a question get it right. On the other hand, because we have a crowd of enthusiastic volunteers at hand, Darryl and his colleagues are freed from trying to do anything especially novel with machine learning. Picking the right machine for the task is important, and so is making sure you understand what it's doing and how it can best be trained, but that's a long way from needing to explore the bleeding edge of the deep-learning revolution.

This approach works well when we're hunting for objects which are relatively common. Supernova hunters should expect to be successful, at least with modern data sets where the telescope and camera are understood well enough to avoid too many false positives sneaking through. But there are plenty of problems in astronomy where a successful end to even a dedicated hunt will be a rarity.

Planet hunting is one example, though here too some judicious filtering can help. But some objects just are intrinsically rare, and will only rarely be stumbled across. Perhaps my favourite of these rarities are gravitational lenses, the result of Einstein's theory of general relativity and a cosmic coincidence.

Gravity, Einstein's theory tells us, is nothing more or less than a geometrical effect. In other words, we feel gravity because of the bending of space by mass. This in turn means that anything passing through space near a massive object will find itself deflected because instead of travelling through flat, empty space it will find itself on a curved trajectory. This rule applies regardless of the mass of the moving object, and even to light. So a key prediction of the theory is that light rays will be bent by passage around a massive object, a fact famously used by Eddington to carry out one of the first serious tests of relativity by recording the positions of stars visible near the disc of the Sun during the total solar eclipse of 1919.

(Two points of pedantry. First, it is possible with some assumptions to derive a light-bending effect from Newton's theory of gravity, and this was done long before Einstein came along. The magnitude of the predicted distortion is different though, and Einstein turns out to be right. Second, there's some modern griping about whether Eddington's results were actually accurate enough, given challenging weather and difficult conditions on his eclipse expedition, to provide a sensible test of relativity. Press coverage from the time, though, shows that whatever the reality this experiment was perceived as important and as elevating Einstein above Newton.)

The idea that our images of distant sources might be distorted by gravity was little more than a curiosity until large and deep surveys of galaxies got going. In just a few places in the Universe, the distribution of galaxies is such that a distant system will lie almost precisely behind another, nearer galaxy or cluster of galaxies as seen from Earth. When that happens, the light from the more distant system will be bent by passing the closer system. The effects depend on the exact geometry. If the alignment is exact, we end up with four identical images of the distant system, one on each side of the nearest system. This is an Einstein cross, and a handful of these remarkable systems are known.

More commonly, the alignment isn't quite right. The more distant object might be slightly displaced from the line of sight, or the internal structure of the nearer object will distort the light. What you see then is a smeared-out image of the distant system, often magnified by the lensing effect of the process. Gravitational lenses like this act as nature's telescopes, allowing us to see distant galaxies which would otherwise be invisible, though as their optics are imperfect the resulting images are distorted.

Even better, their blurry images contain information. The degree of bending of light depends on the amount of mass present in the lens, and on its distribution, and so we get to 'weigh' the objects involved through careful modelling. Sometimes amazing things happen—take the Einstein cross known as MACS J1149.6+2223, which has four images of a galaxy whose light has taken over nine billion years to reach us lensed by a system some four billion light years away. A single supernova has been observed in this galaxy not once but four times, once in each image. In other cases, there are time delays between the appearance of such supernovae caused by the different lengths of the paths that the light in each image takes to reach us.

I find these results astounding. The idea that we can see something that far away, apply knowledge of the Universe and its constituents that is good enough to understand why we see this apparent repetition, and then use that knowledge to understand more, is the kind of thing that got me hooked on astrophysics, and on observational science. Gravitational lenses are amazing, and yet only around a thousand of them are known after years of searching.

It'll be no surprise by now that astronomers want to find more of these things, and that LSST has searching for such lenses as a core part of its programme. It's probably not a surprise either that there's a citizen science project to help, especially as with only a small number of examples available machine learning is going to struggle to help. SpaceWarps, the Zooniverse programme aimed at searching for gravitational lenses, has been hugely successful.

My favourite of its discoveries was found nearly live on TV, as part of a collaboration with Brian Cox, Dara O'Brien, and the team behind their fantastically successful *Stargazing Live* show which once a year takes over prime-time BBC TV for three nights of astronomical chatter. The topics chosen are usually pretty random, but for the last six runs of the programme we've persuaded them to ask their audience to help us with a citizen science project.

The pace of these projects is always exhausting. Television is a strange world, and live television an even stranger one. The programme was based for many years at Jodrell Bank, still home more than sixty years after its foundation to the third-largest steerable radio telescope in the world. A crew of more than fifty people is needed to transform this working observatory into a television studio, with lights and camera needing to be rigged in the most unlikely places before any action can be broadcast to the outside world. Add in the vagaries of the British weather and the logistics become nightmarish.* None of it makes for an ideal

* The most recent BBC *Stargazing* went to Siding Spring Observatory in rural Australia in an effort to escape Manchester weather. It got hit by the tail end of a tropical cyclone.

opportunity to get science done, and the Zooniverse crew usually end up shoved into a corner, craving a decent internet connection to the outside world.

Over the years, thanks to Stargazing Live, we've found planets, studied Mars, and more, but with SpaceWarps we wanted to be still more ambitious—promoting the project on the first night of the show in the hope (and certainly in the expectation from the BBC crew) that we'd find something worth announcing forty-eight hours later. As we set up for that first broadcast, I lost count of the number of people who 'just popped in' to ask whether we were really going to find something.

The chaos doesn't die down immediately after the show. In talking to Brian and Dara I announced the project, and managed to report quickly on the flood of classifications heading our way. In Oxford and in Chicago our team watched as their beautiful infrastructure stumbled under the sheer weight of wannabe scientists before recovering as somewhere in West Virginia servers sent image after image off to eager classifiers. Meanwhile, those of us at Jodrell Bank scrambled to clear the site and head back to the team hotel, leaving the observatory alone.

As a result, it was in the incongruous setting of a conference hotel bar that I found myself staring at a laptop screen bearing what looked for all the world like a neat red lens, an arc of light curving around a nearby galaxy. As producers, presenters, and crew waited for the adrenaline from the night's broadcast to wear off, or huddled in corners to discuss scripts, my Zooniverse colleague Rob Simpson and I stared at the screen. We had something, but we weren't sure what (Plate 10).

It was the red colour that was confusing. Red, in this game, means distant, a sign that the light that the telescope is receiving has been substantially stretched by the expansion of the Universe during its journey from source to us. If this lens was real, it was clearly a distant one, a prize catch, but the colour that made it interesting also meant that we were suspicious of our prize.

We slept on it, but the next morning there wasn't too much more to say. Sipping much-needed coffee, we started the search for previous observations of the new object. It turned up initially in a catalogue called FIRST, a map of the sky as seen by the Very Large Array in New Mexico. Our lens—if it was real—was emitting radio waves, and this was good news. First, it made the thing more interesting; those radio waves must have a source, which meant extreme star formation or an actively growing black hole at its centre, both interesting things in a source as far away as we thought this was. Second, it meant that we could easily design an observation to measure the redshift and hence the distance of the lensed galaxy.

'What we need', I said to Rob, 'is a radio telescope.' He didn't reply, but turned slowly to look out of the window. Staring back at us was the giant dish of the Lovell Telescope. Normally we'd scramble to apply for time, but the Lovell was standing unused thanks to the small matter of a live broadcast happening in front of it. Negotiations followed; Tim O'Brien, the observatory's director, was keen to help, and we eventually persuaded the BBC that they didn't mind if we ruined their carefully planned shot by pointing the telescope away from the studio and towards our target. A few hours later, Rob and I danced in the pouring rain as the floodlit telescope turned slowly on its bearings (repurposed from First World War battleships) to point at a source that had been found less than twenty-four hours earlier.

As ever, observing is only the start of the work, and I will always remain grateful to the Jodrell astronomers who stayed up all night, working on the tricky problem of removing the distinct signature of a live broadcast from the data they received from their radio telescope. It turned out our lens was a broken ring, viewed with light that had taken more than ten billion years to reach us. The red radio ring ended up being the target of observations with telescopes from Hawaii to Mexico. It's magnified by ten times because of the lens, and seems indeed to have an actively growing black hole as well as being a dramatically powerful factory of stars. It is a glimpse of a time when the Universe was at its most active, a time when most stars were being born.

It's also, and to me just as importantly, another example of the ability of citizen scientists to go beyond what they've been taught; despite the fact that all the examples given were blue, the volunteers were able to recognize this red streak as something worth marking. As lenses are rare things, even in the era of large surveys like LSST, we're unlikely to assemble a large training set with which to train a lens-hunting machine; there's progress to be made, perhaps, using training sets of artificial lenses, but for the foreseeable future this will remain a fertile hunting ground for citizen scientists.

What we can do is improve the odds of finding such things. Because the appearance of a lens is shaped by basic laws of gravity, predicting what a lens around a given nearby galaxy will look like is a fairly simple matter. (Well, you'll need a decent computer, but the principles are simple.) That meant that the SpaceWarps team were able to create artificial galaxies to insert into their project. I was a bit worried about this, unsure about how our volunteers would react to being asked to classify 'artificial' data (we were careful not to call them 'fake' lenses).

We needn't have worried. The fact that for these galaxies we knew what the 'right' answer was meant that we could give volunteers feedback, which they craved. While anyone taking part in the project had overcome at least some of the barriers to thinking of themselves as scientists, the odd pop-up confirming that they had the right idea turned out to be extremely welcome.

After all, even the most confident of us need reassurance that we're carrying out a task well every so often.

The real innovation, though, was that we could measure how people were doing. The SpaceWarps team can measure the skill of their volunteers, which they define as the average quantity of information provided by a volunteer presented with a random image from those available to be classified. I'm deviating slightly, deliberately, from the language the project team themselves use here. They're an exceptionally thoughtful and careful bunch, and Phil Marshall in particular—one of the three leading scientists alongside Aprajita Verma and Anupreeta More—is one of the nicest people you could ever hope to meet. As a result, the idea of labelling volunteers, in all their human complexity, with a rating derived from nothing more than a few clicks on a website was anathema to Phil. In all the team's papers, therefore, they set up a system where they represent each volunteer by an 'agent'. An agent is a representation of the volunteer, but necessarily an imperfect one, as the agent knows only about the volunteer's behaviour within the project. We can then label the agent, knowing they are a poor reflection of their human counterpart. I'm less fastidious, and am happy to trust that you know I'm not really reducing people in all their glorious complexity to their performance in one project.

This sort of analysis is useful for checking on the progress of the project; looking at the distribution of skill one sees that the average volunteer is pretty good. While both the highly skilled and the more confused contribute a few classifications, those who go on to contribute tens of thousands of classifications are all highly skilled. This data alone doesn't tell you whether people are learning as they go, so that their skill inevitably improves over time, or whether those who are struggling are simply giving up, but it does show that we're not wasting people's time.

The real power comes when we move beyond this simple, single value. The SpaceWarps model sets up what's known as a confusion matrix for each volunteer, keeping track of four key numbers. For each contributor, we estimate first the probability that they will say that there is a lens when there is indeed one there; second, the probability that they will say there is a lens when there isn't; third, the probability that they'll say there is nothing there when there isn't; and finally (deep breath) the probability that they'll say there's nothing there when there is indeed a lens.

Armed with this information, we can find ways to get more knowledge out of the system. There are four kinds of volunteer to consider. There are those who are always, or nearly always, right; the SpaceWarps team called these 'astute' volunteers, and they are very welcome in any project. There are also those who are always wrong, who miss lenses when they're there and who see them when they're not. These people are just as useful—someone who is wrong all the time provides just as much information as someone who is right all the time, as long as you know that they're wrong.* So because we're able to use the simulations to measure how people are doing, we can increase the amount of useful information we can get from the project.

There are two more categories of people. There are optimists, who see a lens where there isn't one but are reliable when they say there's nothing there, and pessimists, who miss lenses but are accurate when they do identify one. Once we've spotted someone's proclivities, we can work out how seriously to take their opinions, but we can also start to play games with who gets to see what. Before we throw away an image, confident that there's nothing there, then perhaps we should make sure to show it to an optimist,

* You may find this a useful strategy for life in general.

just in case. If we think we've found a lens, then we should show it to a pessimist—if even they reckon there's something there, then our confidence should grow sky high that we're on to something. Playing with task assignment in this way promises much more efficient classification, and more science produced more quickly.

The only trouble is that this gets complicated fast. With tens of thousands of people participating in even a small project, and hundreds of thousands of images to view, the number of possible solutions is unbelievably large. Even when we consider that our choice is restricted by the fact that not everyone is online at the same time, complex mathematics is required to work out what a sensible path is. Work by Edwin Simpson of Oxford University's Department of Engineering showed quickly that clever task assignment could produce results of the same accuracy with nearly one-tenth of the classifications, an enormous acceleration and one that is especially welcome when looking for the rarest of objects.

SpaceWarps is among the most sophisticated Zooniverse projects in how it treats its data, and in offering a faster route to science it seemed to be a template which we could apply in all of the other fields that we're working on. Plenty of work on this sort of task assignment has been done by researchers in a field of computer science known as human–computer interaction, typically using Amazon's Mechanical Turk system to connect researchers with those who will complete tasks for small payments.

Yet things aren't so easy with citizen scientists who are themselves volunteers, and a simple experiment with a project we ran called Snapshot Serengeti shows why. Whenever I lecture on the Zooniverse, one of the most common questions is whether we really need humans given all the progress in machine learning. I've hopefully dealt with this already, but the disease seems especially acute around projects like this one, which uses motion sensitive cameras to monitor wildlife in the Serengeti National Park.

The images the cameras produce are wonderful, beautiful, and varied. Some would easily grace the cover of National Geographic, while others are more quirky. The team's favourite comes from a camera programmed to take three photos in quick succession once triggered. The first of this particular sequence shows a hyena staring at the camera as the flash goes off. The second shows the same hyena skulking innocently in the background, but the third shows some sharp canines and the inside of the hyena's mouth. Apparently getting chewed by the local wildlife was a common end for the project's cameras (not a problem Penguin Watch faced in Antarctica), and elephants using camera stands as scratching posts didn't help either.

Despite the immense variety in what the project's cameras capture, there seems to be something about the task of identifying animals in images that seems to convince people they can quickly write a script or produce an off-the-shelf machine-learning solution that will solve the problem. It turns out it's harder than it looks. While we'll share our data with anyone who wants it, no one's yet come up with a completely robust solution yet. I have a soft spot for the attempts of a team we worked with at the Fraunhofer Institute in Munich (home to the inventors of the MP3, the format which encodes music on your phone and other digital devices) who developed an especial dislike of ostriches, which thanks to their bendy necks and bandy legs turn out to be able to twist into a computer-defying set of shapes.

Nonetheless, some tasks are definitely easier than others. Wildebeest are common enough to trigger complaints from regular classifiers, and so building up a suitable training set for them will be easier than, for example, doing so for the small, skunk-like zorillas which appear in one in every three million images (Figure 22). Easiest of all, though, is to identify the images

**Figure 22** A rare image of a zorilla as captured by the Snapshot Serengeti cameras.

with precisely zero animals in them at all.* Almost three-quarters of the data consisted of such images; either a camera would malfunction, and take image after image of nothing until its memory card was used up, or waving grass would do a good enough impression of a passing lion that it too would be captured.

We know that volunteers care about getting science done, and we hate wasting their time, so removing these animal-free images was an obvious thing to do. What happened next was surprising. As volunteers saw more and more images with animals in, the total number of classifications the project received dropped. People might like contributing to science, but in trying to make

* Notice I do not, as I would have done once upon a time, call these 'blank' images. I was cured of that when speaking to a room full of plant scientists. Pointing at an image of a tree and grassland, I confidently told them there was 'nothing there' and saw the audience rise up as one. Apparently they call it plant blindness.

it faster for them to do so we'd done something that made the experience less pleasing, and we weren't quite sure what it was.

One theory suggested that there was a total amount of work that people would be willing to invest in the project. It's faster and easier to say that there is nothing in an image than it is to distinguish a Thompson's from a Grant's gazelle, and so maybe by giving them more to do we were using up people's effort faster. I don't think that's the right explanation; we know that all else being equal encountering an animal in Snapshot Serengeti made people more, not less likely to keep classifying, and so it seems to me that we would be at least as likely to encourage as discourage people from classifying.

Instead, I think we'd changed how exciting the project seemed to people. Whereas before they'd seen nothing, then nothing, then nothing again, nothing again, and then suddenly a zebra, now they endured the apparent tedium of zebra followed by zebra followed by wildebeest followed by yet another zebra. While almost all the research on how to assign tasks for efficiency uses paid subjects, who can be assumed to stay put regardless, our volunteers are free to walk away at any point. By trying to make things better, we made their experience worse. The choice between getting more science done and providing 'fun' online is stark, even with such a simple experiment.

This, of course, won't be a surprise to any game designers who are reading. Since the first computer games bleeped their way into our collective consciousness in the 1970s and 1980s, players have been participating in an enormous collective experiment to find what will keep us clicking. While almost all games pay attention to this, it's most obvious in simple phone games that occupy so many commutes, most of which are optimized to produce just the right level of micro-excitement to keep us clicking. I don't mean to sound snobbish about this, not least because I'm currently about 500 levels into something called *Two Dots*. We, as humans, are just wired to respond in this way, and we behave as if we're playing games even when it's not deliberate. In the early days of Galaxy Zoo, lots of people told us that their experience of classifying galaxies was like eating crisps; you don't mean to have just one more, but you do, again and again and again, rewarded with the next image each time you click on a galaxy.

The implications of this seem obvious. For all that Zooniverse projects are scientific projects, they are also experienced as games. We could, perhaps, make them much more popular by manipulating the data so that a suitable fraction of animal-free images were served without worrying about whether such classifications were useful. We might take the most spectacular images and make them appear more frequently, even if we already know what they show, making further classification redundant. If manipulating the data this way makes for more classifications and hence more science overall, then perhaps there's no harm.

Plenty of projects have taken this route, and walked much further along it than we have. It feels like an obvious choice. If people like playing games, and are willing to contribute their time to do science, then a game that lets you contribute to science feels like the best of both worlds. But this feels like a step too far for me. Our participants take part because they want to contribute to science; it feels wrong to feed them images that we don't need help with. This kind of dilemma will only become more acute once machines start picking up more of the slack, and we start deciding what is really worth sending to classifiers.

At the end of the book, I want to use these ideas to talk about where citizen science is going. First, though, I need to tell you about what has clearly become the real strength of public participation in Zooniverse projects—the ability to find the truly unexpected, and to uncover stories of objects which would otherwise remain forever hidden from view.

# 7
# SERENDIPITY

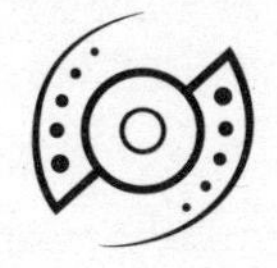

In SpaceWarps and other projects, it's clear that people, unlike machines, cope well with the unexpected. The example of the red-lensed galaxy shows that nicely, but it's a risky argument. As training sets become larger, it's going to become harder to surprise a machine, and so taking this to its logical conclusion one's left with a vision of the citizen scientists of the next decade being chased from task to task as machines improve. Hunting the rarest of objects might still be a useful occupation, but the opportunities to make a real contribution will become scarce. Given all the good that comes from projects that offer everyone a chance to help science, that would be a shame.

It's premature, I believe, to declare Zooniverse-style citizen science a passing phase. There is a more interesting future in store—one in which the line between the work done by amateurs and professionals, and between the amateurs and the professionals themselves, blurs still further. Evidence for this future is found in stories from many projects, in discussions that spring up around the unusual and the unexpected. I could give many examples, but let me tell you about two that I was personally close to. They're stories of old-fashioned science, in which professional and citizen astronomers used a bucketload of ingenuity to work out the solutions to new mysteries. These stories involve groups of people from a variety of backgrounds and with myriad life experiences.

The first story dates back to the crazy first year of running Galaxy Zoo. The forum had quickly become a busy place, with posts about anything from astrophysical techniques to tea, but among the creativity of that community there was plenty of chat about what people were seeing on the site. A Dutch school-teacher named Hanny van Arkel was the first to point to a blue blob that appeared near an otherwise unremarkable galaxy in one of the images.

The galaxy had a catalogue number—IC2497—and Hanny named the blob the 'Voorwerp' (Plate 11). When, a little later, the Galaxy Zoo team found her posts, I think we all assumed that it was a Dutch technical term. It turns out to mean 'object', or 'thingy', but 'Hanny's Voorwerp' is now the official name of the blob, endorsed by several major journals. To be honest, at the start the most interesting thing about the Voorwerp was probably the amusing story of the name, but Hanny wanted to know what it was.

If I'd come across the blob while sorting through images myself, I think I'd have ignored it, placing it to one side while getting on with more straightforward tasks, if I would even have noticed it at all. Yet Hanny, the citizen scientist, was captivated by the discovery and pressed us to find out more. It was an early lesson that experts aren't always right; not only do highly trained professionals occasionally make silly mistakes, they also can't always be trusted to focus on what is truly interesting.

There are plenty of examples scattered across the scientific literature. Take the work of a group led by Trafton Drew at Harvard Medical School, for example, presented in the journal *Psychological Sciences* with an arresting title straight from a horror film: 'The invisible gorilla strikes again'. (Astronomers need to have more fun with paper titles.)

The invisible gorilla teaches us not only that experts make mistakes, but that they're more likely to do so than the rest of us in some circumstances. The gorilla in question is a small cartoon figure, posed with one fist in the air for reasons known only to itself. It was placed by Drew's team into images produced by CT scans of patients' lungs, grainy black and white images studied by surgeons to look for signs of cancer. The participants were medics on the look-out for anomalies; an ideal, expert crowd for gorilla spotting. To make the task easier, the researchers made the gorilla larger, by a factor of nearly fifty, than the cancerous nodules the researchers were supposed to be looking for. Frankly, unless they'd equipped the beast with a party hat and balloons it's hard to imagine how they could have made it more obvious.

The results are shocking. Of the twenty-four experts who took part in the challenge, twenty of them missed the gorilla completely (Figure 23). They didn't see it. They didn't mistake it for anything else—how could they?—but their brains just didn't register something they weren't expecting. When I first heard about this, I assumed they weren't trying very hard, but eye-tracking equipment used in the lab showed that most of those who missed the simian interloper looked straight at it. Not an absence of effort, then, or a sloppy inspection, but an absence of conscious attention.

Surprising though it is, that's the result that the researchers expected. A previous result had invoked the invisible gorilla, this time wandering among players on a basketball court. If you watch the video, the figure in a party-store gorilla suit couldn't be more obvious, but an audience told to count the number of passes will miss him, even as he pauses to wave to the camera and hence to the inattentive viewer, who remains oblivious.

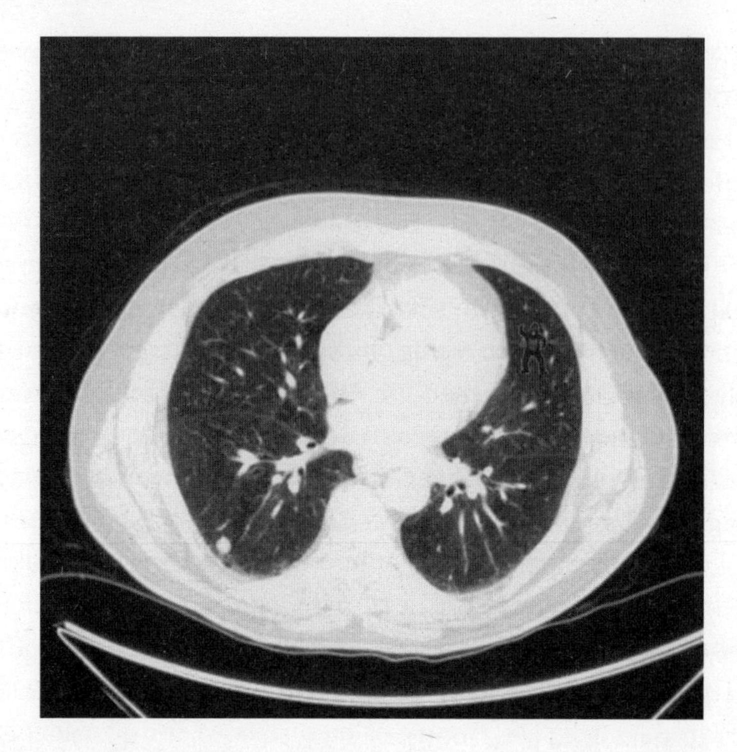

**Figure 23** The invisible gorilla (top right) as presented to surgeons for classification. It wasn't noticed by most of the experts looking for tumours in these images.

Once you know there's a gorilla in shot, it's literally impossible to miss him. So famous has the experiment become that for years it ran in cinemas as a road safety advert, preaching the need for careful attention. Yet not everyone is fooled to keeping their eye on the ball; expert basketball players are much more likely to notice the gorilla in their midst.

It's not hard to explain why experts perform better. If you're more used to following the movements of an orange ball whanged around a court by a bunch of players wearing vests, I'm willing to bet you're also looking for different things from the rest of us. Those who know basketball will, I reckon, be looking not at the ball but for people to pass to, and will therefore instantly be aware of the offensive threat posed by a gorilla near the three-point line. Expertise here involves carrying out the simple task of counting passes automatically, such that it ceases to consume effort, freeing you to look beyond the simple task and see the whole game.

That's why the CT scan study is so surprising to me. Here, even experts can't be trusted; what the study is recording is a phenomenon known as inattentional bias, and it afflicts us worst when the object being searched for—cancer nodules in this case—is very different from the interloper. It's harder to spot things the more different they are from the things you're looking for. The gorilla being obviously not a nodule doesn't help, but rather ensures that the nodule-searching brain dismisses it before the conscious brain can be surprised by it. Another problem is what people who study this stuff call 'satisfaction of search', the human tendency to stop looking once we've found something. Gorillas close to nodules were spotted slightly more often, but still missed more than two-thirds of the time.

So finding what you're not looking for turns out to be extremely hard, and that has consequences in the real world. Kenny Conley was a police officer in Boston, and at two in the morning on 25 January 1995 he was in hot pursuit of a suspect. An undercover officer was also present, but when they got to the scene Conley's fellow officers mistook the disguised cop for the suspect. They proceeded to badly beat him up, which eventually ended them in deep (and, I reckon, deserved) trouble.

Conley, chasing the real suspect, had run straight past the place where the assault of the undercover officer was taking place, but claimed that he hadn't seen anything at all. No one was able to believe he could have missed what was happening, and he was found guilty of perjury in lying to protect his fellow officers and sentenced to nearly three years in jail as a result.

This case attracted attention, and inspired an experiment. Participants were asked to run after someone, passing on the way a group of brawling actors. At night, only a third of them noticed the fight, and even in broad daylight a third missed noticing that anything was happening. The act of concentrating on chasing someone decreases the attention one pays to the surrounding world. As the researchers put it in the title of their paper, 'You do not talk about Fight Club if you do not notice Fight Club'. (Psychologists really, really have more fun with their paper titles than astronomers do.)

Conley eventually won an appeal, though not because of the research described above. Nonetheless, once you start thinking like this it becomes obvious why Hanny, and not a bunch of astronomers, found the Voorwerp. Expert—and especially professional—classifiers know what a galaxy looks like, and so aren't likely to be distracted by the appearance of something else. Newer volunteers, those with less knowledge, are likely to be conscious of all sorts of things in the data, some interesting and novel and some not.

Hanny was also an effective advocate for her discovery. It was her Voorwerp, after all, and she wanted to know what it was. Her desire to understand pushed us on the Galaxy Zoo team to look into the matter, but it wasn't easy at first. A leading hypothesis at first was that it might have been an interloper, a nebula or cloud of gas belonging to our own Milky Way. In order to test this idea, we needed a spectrum of the object, but the Sloan Digital Sky Survey that provided the image in which it was found hadn't targeted it for spectrographic follow-up. The survey's algorithms just hadn't anticipated anyone finding it interesting. Worse, the thing was faint enough that we needed a large telescope to get enough light, but time on those is won by writing convincing pitches describing future scientific bounty that will inevitably flow from a particular set of observations, not from following some wild goose chase inspired by a single image of a weird object.

If it were up to me, we'd be able to write 'We found an unusual thing and want to look at it' and send that off to the Time Allocation Committee (TAC). A TAC is not some sort of committee of advanced aliens, something from the Doctor Who cutting room, but rather the group convened by an observatory that decides who gets time using the telescopes. Faced with more astronomers with more ideas than anyone should have to deal with, they tend to look askance at speculative proposals.

Luckily, astronomy is a small world. I found out that Matt Jarvis—both then and now again a colleague in Oxford—happened to be observing for his own purposes at a telescope in the Canary Islands. With a little nudge, Matt pointed the telescope in the right direction, and emailed back a spectrum. Whether Matt had sacrificed his own observing time, or 'accidentally' pointed the telescope at the Voorwerp while setting up for the night, I've never dared ask.

There's always been a somewhat informal barter economy around telescope time; emails soliciting objects that would be worth observing after primary targets were set, or phone calls requesting emergency—or risky—observations, used to be common. As scheduling has become automated and efficient, these loopholes are closing. I worry about how we're going to take risks, and think most observatories should set aside a small amount of time for observations which might be a bit unusual but which might pay off spectacularly.

Anyway, we got our data and the spectrum was a revelation. Even a quick glance at it told me that the Voorwerp—whatever it might be—was at more or less the same distance as the neighbouring galaxy. It was therefore huge—almost galaxy-sized itself.

Looking back, I think that was the first moment I realized this was more than a curiosity; that the Voorwerp wasn't just something that had caught Hanny's eye, but was genuinely interesting in itself.

It clearly needed more than a casual glance. Luckily, I was sitting in the conference centre in the middle of Austin, Texas, at the winter meeting of the American Astronomical Society—the largest annual gathering of astronomers. Among them was Alabama professor and Galaxy Zoo observing guru Bill Keel, who quite literally wrote the book on how to study galaxies. Dealing calmly with me waving a laptop in his face, Bill immediately noticed what I had not; there were features in the spectrum which suggested the presence of elements such as sulfur,* in conditions which led their atoms to be highly excited. In other words, the gas in the Voorwerp was hot. Very hot. About 50,000 degrees Celsius in fact, or nearly ten times the temperature of the surface of the Sun.

That's not unprecedented, but it does need explanation, all the more so because there was nothing in the spectrum which suggested the presence of stars embedded in the gas. If they'd been there, they would have contributed what is called continuum light, shining at all wavelengths, but the absence of a significant continuum meant that very few bright stars could be present. There certainly weren't enough to heat the rest of the gas. Sitting in the corner of a corridor in a large and almost completely soulless convention centre, Bill and I realized the Voorwerp was a real mystery. Understanding why this blob of gas was excited, and identifying the source of its excitement, was a proper scientific question, and the spectrum Matt and colleagues

* I'm using the American 'sulfur' not the English 'sulphur' because that's what the International Union of Pure and Applied Chemistry says we should do. They adopted English spellings of aluminium and caesium, so it's not as if the Americans got everything their own way.

provided was not the end but the beginning of a scientific detective story. With an unusual spectrum in hand, we had the ammunition to go and chase down more clues.

First, though, there was a chance for some good old-fashioned speculation as we tried to work out what sort of thing the Voorwerp might be. One obvious possibility was that it was the remnant of a supernova; many of the explosions discussed earlier in the book will produce not only a dense remnant—the neutron star or black hole that forms from the star's core—but also a surrounding cloud of gas. These supernova remnants don't last for long—we're watching the one produced by the 1987 explosion in the Large Magellanic Cloud change before our eyes—but they do shine brightly due to gas excited by the explosion. The shock wave from such an explosion might, if powerful enough, excite surrounding gas to the degree seen in the Voorwerp. A careful look at the Voorwerp itself supported this nascent theory; there's a roughly circular 'hole' in the gas which would easily be explained if it was centred on the site of an explosion, with the gas closest to the action having been destroyed or ejected completely.

It's a simple, neat explanation of the object, which is utterly confounded by the facts. The biggest and most immediate problem is the sheer size of the Voorwerp itself, much, much larger than any supernova remnant in the Milky Way. Any explosion capable of exciting gas over such a large volume of space must have been quite something, but in the early speculative phase of thinking about things we weren't too discouraged, given freedom to imagine the unlikely because the Voorwerp was, as far as we knew, one of a kind.

The discovery of many such objects would mean making a claim about how frequent supernovae capable of producing such massive remnants are, a calculation that could be quickly tested by observation. With only one example, who's to say we hadn't stumbled across the remains of the most super of supernovae? The next stage though, is to calculate, or at least guess, how long the unique thing you're observing will survive in something like its current state. If it's short lived—a nova that will come and go in a matter of weeks or a planetary nebula which will last for only a few tens of thousands of years—then an argument which relies on scarcity is dead in the water; you may only see one example now, but another will be along in a little while. If it's long-lived—and the sheer size of the Voorwerp, closer to the scale of a dwarf galaxy than any normal remnant, suggested it wasn't going anywhere in a hurry—then the argument that you might be dealing with an exceptional specimen has more weight.

I'm labouring the point perhaps, but this sort of argument lies right at the heart of the kind of astronomy I like to do, and what, when whiling away nights back in the school observatory, I thought astronomers mostly did. We found a weird thing. Great! How weird is it? Is it especially close, or far away? How bright does it look? Is it changing? How does it compare to other things? These are all simple observations, but they're as much part of attempting to understand the Voorwerp as writing down equations that convert features observed in a spectrum to physical properties like temperature.

In this case, the line of reasoning suggested that any Voorwerp-producing supernova would have to be exceptional, and therefore exciting. Before we could go searching for the dense remnant, the neutron star or black hole that would confirm that a giant explosion occurred here, we realized a clue had been overlooked. The Voorwerp is large enough that we don't need to treat it all as a single object—we can look at parts of it separately, even from our distance of three hundred million light years.

Once we realized that, a pattern became clear. How excited the gas was depended on how far it was from the neighbouring galaxy, with the gas closest to the galaxy less excited than that which is further away. A small clue, but an important one. If this was a hard-boiled detective movie, imagine the camera panning slowly to the galaxy, ominously hanging 'above' the Voorwerp in each of the images we'd spent ages staring at.

Could it have been the culprit? Like most, and probably all, large galaxies, it contains a supermassive black hole at its centre. As material falls into such a black hole, if there is enough of it, it can form a disc of material orbiting the central black hole, known as an accretion disc. The physics of how material behaves in such circumstances is, well, complicated to say the least, but it's clear that such systems can produce powerful jets of material moving at high speed. Such jets are common in massive elliptical galaxies—M87, the giant at the heart of the nearby Virgo Cluster, has a well-studied example which moves at very nearly the speed of light—but they have also been found in spirals. Volunteers in the Radio Galaxy Zoo project, which tries to pair galaxies observed in the radio with their counterparts in the infrared, have found just such an object, and in that case as in almost all spiral systems with jets, the jet was perpendicular to the disc.

So could such a jet, produced by activity in the core of IC2497, have excited the glowing gas in the Voorwerp? Such a scenario seemed more plausible after a close look at the neighbouring galaxy itself. It seemed to be warped, its disc twisted and marked by thick dust lanes—features which suggest a recent interaction with a neighbour. Such an interaction, or even a merger, might have funnelled material down into the galaxy's central region where the black hole lurks, piling on material and making the presence of a jet much more likely, and if such a jet existed at right angles to the galaxy's disc, it looked plausible that it would hit the Voorwerp directly.

Mystery solved. Yet the details didn't add up. If there was enough material falling into the black hole to produce a jet able to excite the Voorwerp, then the gas and the dust that immediately surrounds the black hole should also be warm, glowing brightly in the infrared. The Voorwerp's neighbour is bright in the infrared—it's what astronomers call an LIRG, or luminous infrared galaxy—but it wasn't bright enough to allow for the necessary activity. Despite the neatness of the explanation, it was clear that we couldn't hide a powerful enough source of activity at the core of the galaxy to account for the Voorwerp's brightness.

This was confusing, but slowly, in discussions and emails and phone calls, those of us on the Galaxy Zoo team who were increasingly consumed with trying to answer Hanny's simple question realized that we had to consider the possibility that the Voorwerp was an evolving object, capable of change. The distance between the Voorwerp and the galaxy is about 50,000 light years.* This means that light travelling from the centre of the galaxy to the Voorwerp, and then heading to Earth, would arrive 50,000 years after light taking the direct route. Put another way, the Voorwerp is a light echo—it reveals to us the state of its neighbour galaxy as it was 50,000 years ago.

The fact that the Voorwerp is highly excited now tells us that, back in the stone age here on Earth, the black hole at the centre of IC2497 would have been feeding heartily, even though today it is relatively quiet. Had our Cro-Magnon ancestors had stone age binoculars, they would have found the galaxy—now too faint to be seen with small telescopes—easily visible, the nearest example of what we call a quasar. In the time that's passed since then—

* That's the projected distance on the sky—the distance you get if you assume that the lines joining us to the Voorwerp and the Voorwerp to the galaxy make a right angle. If the Voorwerp is in front of or behind the galaxy, that distance might really be only 25,000 light years or as much as 75,000 light years.

not long in the 13.8 billion year long history of the Universe—the galaxy must have quietened down. Perhaps the black hole ran out of fuel, or the activity around the black hole might have heated the surrounding gas, preventing further collapse in a process known as feedback. In either case, as long as something dramatic had changed in the galaxy in the last few tens of thousands of years, it would explain what we saw.

This is an exciting idea. We can monitor what the black holes at the centres of galaxies are up to on human timescales, returning year after year to the same objects to see what, if anything, has changed. We can see too how the population of active galaxies changes over cosmic time, by comparing the population of galaxies that we see at different epochs of the cosmos's history. But being able to trace changes on a timescale of thousands of years, in-between these two more accessible timescales, is unique.

Not that having had the idea was enough. Getting from there to a proper result was hard work, and a particularly hard slog for me; my PhD had made me an expert in using radio telescopes, not in handling data from anything with a mirror, and so I was learning what to look for in spectra and in the other data we gathered as we went along. With much help from Bill and others, eventually a short paper was ready, and we sent it off to our normal research journal.

We didn't have to wait long for the referee's report to appear in my inbox. Unfortunately, it was as critical as anything I'd seen. My analysis had, it said, basic problems, and the interpretation I've just spent paragraphs convincing you of was overblown. Our conclusion that the black hole couldn't currently be active enough to account for the Voorwerp's observed brightness was wrong, we were told.

In some fields, such a bad report would have meant the end of the paper's chances of publication. In astrophysics, there's a more collaborative approach and it's possible to go back and forth with the referee until all is well, or until mutual exhaustion ends the process, at which point the editor as umpire steps in and ends the fight. This tangle with the referee was one of the most exhausting of my career, and we went back and forth more than a dozen times; what had started as a four page paper weighed in at thirty pages by the time we had satisfied every niggling doubt in their mind (or, I suppose, won through sheer exhaustion).

This effort mattered. One reason for publishing in a journal is to show publicly that a result stands up, that it has at least passed the minimum standard required for peer review. A second might be to spread the word, communicating a result to one's colleagues. Another goal is to build up a public track record of work—even though there are well-documented problems with the approach, we use someone's publication record when considering them for jobs or for promotions. The fourth reason, though, is to stake a claim for posterity. Despite all the wrangling and the back and forth, we needed the paper out so that Hanny—as a co-author alongside the rest of us—could get the proper scientific credit for her discovery.

The trouble was, the Voorwerp was big news. Hanny had been on Dutch television, who were reporting on the mysterious blob, and we'd been blogging to share our progress and our excitement with the Galaxy Zoo volunteers. As a result, a Dutch team of astronomers had decided to take a look at the new object themselves, using a network of radio telescopes to get a very different view of the system, one more detailed than had been obtained at similar wavelengths before. Before our referee was happy and our paper could be accepted, this rival group published their own paper which told a different story.*

* This Dutch group were kind enough to add Hanny and a few of the Galaxy Zoo team to the author list for the paper, but it was still annoying not to get there first.

A quick glance at their data showed that the Voorwerp was bigger than we thought. It turns out that only part of it has been excited enough to glow brightly when viewed in visible light; there exists a much longer streamer of cold gas wrapped round the galaxy, exactly as one might expect in the aftermath of a merger. The Milky Way's satellites, the large and small Magellanic Clouds, have been disrupted by the encounter with our larger galaxy, and trail gas and stars behind them as a result. The consequences of a more dramatic merger should be even more spectacular, producing long 'tidal tails', streamers of gas that here happen to be in the right place to be excited. Was the Voorwerp just part of such a tidal tail, illuminated by a still-active jet?

The cause of that excitation seemed to be visible for the first time in the Dutch radio data. Deep in the heart of IC2497, it turns out there is indeed a jet of material, moving fast and heading straight for the Voorwerp. This is exactly what we'd originally expected, and completely consistent with a currently active galaxy. By the time our paper was published we had to argue that there were two possibilities—we were either right, and the galaxy was now quiet, or our calculations were wrong and the jet discovered by the Dutch team was evidence that we'd made a mistake.

Luckily, there was a clear test available. Kevin—he of the original classifications that inspired Galaxy Zoo—and friends applied for and won time to use two telescopes in space. *XMM-Newton* and *Suzaku* are European and Japanese telescopes that look for x-rays, high-energy radiation emitted most commonly by very hot gas such as that swirling around an active black hole which is still growing. Using x-rays also allows us to peer through the dust clouds that surround the centre of a galaxy like IC2497, getting directly to the heart of the action.

When we used these telescopes to look at the centre of the Voorwerp's neighbour we saw precisely nothing. There were a few stray photons recorded by the detectors, but nothing worth writing about. In other words, the galaxy was dead, just as we had predicted. And the implications are startling. The black hole at the centre of IC2497 probably weighs in at a few million solar masses, and yet a system dominated by an object that large has managed to change its behaviour completely in the small matter of a few tens of thousands of years.

In the past, the Voorwerp must have been the brightest quasar in the sky, and as it would be too much of a coincidence to have the nearest such object behave oddly, such behaviour must be relatively common. Much better to assume that such things happen all the time than to argue that the nearest such galaxy just happened to misbehave shortly before we came along to watch. The lesson of the Voorwerp is that galaxies switch from active to passive—and presumably back the other way—all the time. Classifying a galaxy as either active or quiescent becomes not a property of the galaxy, like its mass, but a statement about what stage of its life a galaxy is in.

Even the story of our own Milky Way changes when you start thinking like this. The centre of our galaxy is a quiet place today, containing a supermassive black hole but one that is quiet and devoid of any substantial accretion disc. There was great excitement a few years ago when what appeared to be a gas cloud of about the same mass as Jupiter, named G2, appeared likely to fall in, and coffin-chasing astronomers were keen to watch. As it turned out, G2 survived its close passage around the black hole (it didn't come close enough to pass the event horizon, the point of no return) and may well be something more substantial than just a cloud of gas.* Just the fact that everyone got so excited

* The most favoured hypothesis seems to be a pair of stars embedded in a cloud of gas, for a variety of complicated reasons.

about it, though, tells you that not much happens in the centre of the Milky Way.

Look away from the disc of the galaxy with the right eyes and the picture changes. NASA's satellite *Fermi* has been mapping the sky in light that is even more energetic than the x-rays that betrayed the Voorwerp's secrets. The gamma-ray sky is marked by two large bubbles of hot gas, extending symmetrically for tens of thousands of light years and centred on the galaxy's heart. A good explanation for these structures, now known as the 'Fermi bubbles', is that they're a pair of shock waves exciting the thin gas that exists beyond our disc, the echoes of a time not that long ago when the centre of our galaxy was home to an active black hole. As material fell on the Milky Way's central engine, it shone so brightly that as its radiation travels out into space it can still excite its surroundings.

So even the Milky Way can be active, and the Voorwerp helps us understand this behaviour. We also managed to get *Hubble Space Telescope* images and data, which made the Voorwerp yet more famous. (It made a brief appearance in one of David Letterman's monologues, being revealed at the end as nothing more than a smear on a camera lens, easily wiped away.) In the meantime, Galaxy Zoo volunteers in search of their own discoveries quickly assembled a set of 'Voorwerpjes' (Figure 24).* Each of them is a glowing blob of hot gas, excited by activity around a galaxy's central black hole.

The variety of shapes seen in the Voorwerpje sample, many of which have now been imaged by *Hubble* too, is remarkable. There are long braids of gas, which seem to twist round each other.

* Voorwerpen would be the Dutch plural, but these are smaller versions, so we use Voorwerpjes—the diminutive. It's amazing what you end up having to know in this job.

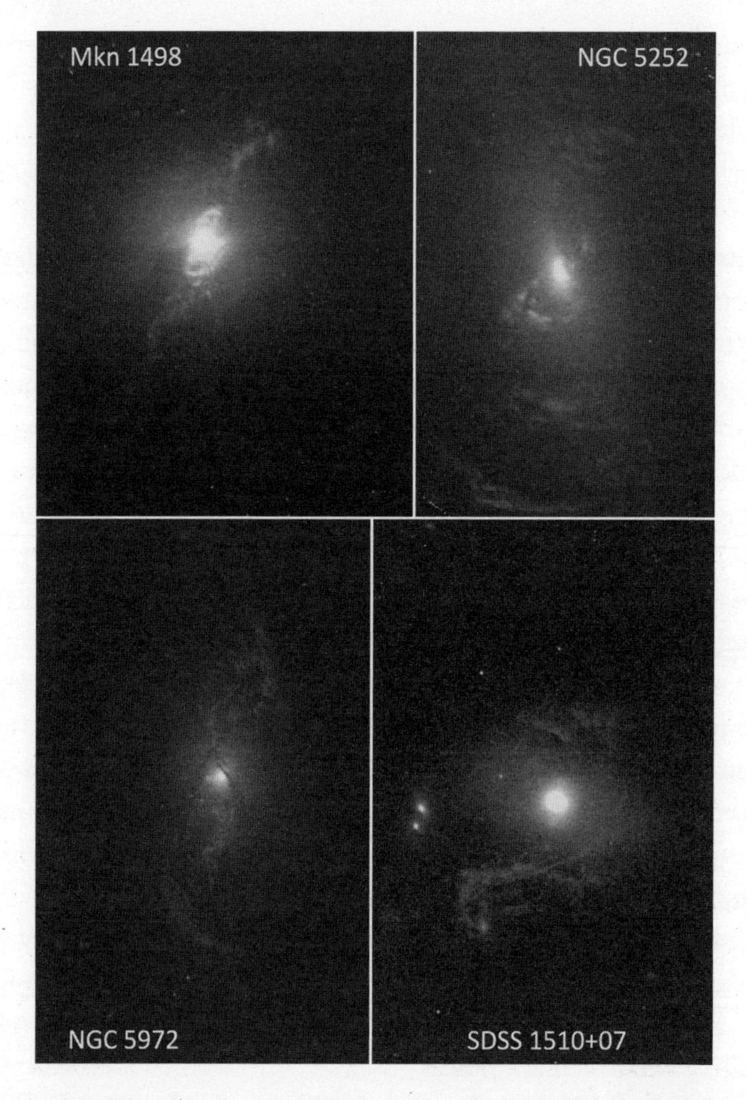


**Figure 24** Example Voorwerpjes imaged with the *Hubble Space Telescope*. The emission is from excited oxygen and shows a remarkable range of shapes, which are difficult to explain.

There are dense clouds, in front of, behind, and surrounding their host galaxies. There are a surprising number of rings, and features that look rather like the still-mysterious 'hole' in the Voorwerp. The complexities of these shapes are either caused by an uneven distribution of gas or rapid changes in the luminosity of material falling down into the black hole, and they don't make understanding these systems easy. Nonetheless, after a lot of work it seems clear that about a third of galaxies with visible Voorwerpen* have faded dramatically, just like the original.

Hanny's find is still inspiring new scientific discoveries a decade after it was found, but it can seem like an sideshow compared to the main Galaxy Zoo story. Galaxy Zoo wasn't set up to find giant glowing gas clouds, nor to reward volunteers who got interested in curiosities they came across while classifying. The vast majority of the scientific papers which have been published as a result of the project use the results produced by volunteers clicking away on the main interface, few of whom get the recognition and dose of fame meted out to Hanny, and pay no attention to random discoveries made and discussed on the forums.

Its importance is rather that it turned out to be the first of many similarly serendipitous discoveries, and in many cases the volunteers went far beyond just pointing at the presence of an object. The idea that citizen scientists could do more than just spotting odd stuff first became obvious when we came across the Green Pea galaxies—a good few months after our volunteers had started thinking about them.

As the name suggests, these are small, round, and green objects which appear in the background of some of the images from the Sloan Digital Sky Survey that populated the original Galaxy Zoo. A small group of volunteers, calling themselves the Peas Corps,† set out to collect them but also, importantly, to find out what they were. First they noticed that the peas were all the same

* Got the plural in too!

† It took me months to realize this was a joke. The revelation finally came when I was on stage, giving a lecture, and said the name out loud.

**Figure 25** A 'pea' as found by Galaxy Zoo volunteers. This tiny galaxy is undergoing a dramatic burst of star formation; most of the light we see is due to emission from excited oxygen.

colour, a blueish-green; and then that the only other things this colour were small, irregular galaxies. Remarkably, the Peas Corp organized their own version of Galaxy Zoo to sort through the 15,000 or so objects that shared this hue, emerging with a set of a few thousand peas (Figure 25). Sloan also provided spectra for these objects, and the volunteers went and grabbed this extra data. When they inspected their haul, they noticed that their light is dominated by emission from oxygen.

The significance of that discovery wasn't immediately apparent, but they quickly decided that having a strong oxygen line was a good test of whether something was a 'pea' or not. Having tested their sample, they sent an email announcing to the Galaxy Zoo team—their professional counterparts—that they'd discovered not an odd object or a new galaxy, but a new type of galaxy.

The bright oxygen line explains the distinctive colour of the peas, and it's a sign of rapid star formation. A quick look at the data available for the sample the Peas Corps had assembled was shocking. The peas were systems in which stars were forming at a prodigious rate; despite being dwarf galaxies with only about a tenth of the mass of the Milky Way, they match or even exceed our own galaxy's production of stars. These tiny systems are the most efficient star factories in the local Universe. Located in the backwaters of the cosmos, in the least populated parts of the Universe, something has caused them to convert all of their gas to stars. Understanding why this is happening, and what it tells us about the histories of their larger counterparts, is a matter of quite some debate, with more than fifty papers devoted to their properties and their nature in the literature.

One exciting idea is that these are the last galaxies to undergo the kind of star formation episode which the most massive galaxies might have experienced early on. Such enormous bursts of star formation might have been responsible for what's called reionization—the point in the Universe's history where neutral gas throughout space was excited for the first time by light from newly formed stars. The peas would then represent our only chance to see what a galaxy undergoing such an event looks like locally.

Unlike the Voorwerp, which to the best of our knowledge had never been spotted before, the peas weren't completely new to science. I've found them lurking in papers going back as far as the 1950s, hidden within catalogues of systems which shine brightly in oxygen and other atomic emission lines. But no one had looked—not taken so much as an idle glance—at these systems, and so no one had noticed that these things don't seem to be normal galaxies and might therefore be worth some attention. It's not even true that only people—citizen scientists—could have found them, as a careful selection of systems with particular colours can lead to a sample consisting solely of peas and not much else, but we needed the volunteers to point us to what was worth selecting before such an exercise would seem worthwhile. Even in modern astrophysics, dominated by big data and machine learning, the critical insight that something might be worth following up remains a very human one.

# 8

# IS IT ALIENS?

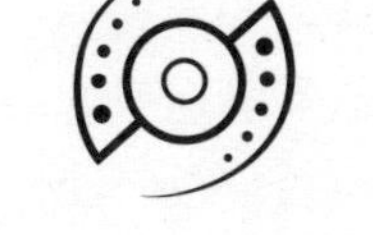

The discovery of wonderful, unexpected things isn't limited to Galaxy Zoo. Volunteers on the Planet Hunters project may—may!—have found the signature of an advanced alien civilization orbiting a nearby star. While perhaps not the most probable explanation for what they've found, the theory is plausible enough for one of the top astronomical journals to publish a paper which seriously suggests that what was found qualified as an 'alien megastructure', words which, as it turns out, get you noticed.

The star in question was once known only as KIC8462852. KIC stands for Kepler Input Catalogue, which tells you that this particular star lies in a particular patch of sky. Lying across the border of the constellations of Cygnus and Lyra, it was stared at by a space telescope called *Kepler*, built by NASA to hunt for exoplanets. *Kepler* is not a large telescope, and it didn't have an exciting task; for three years it stared at that same patch of sky, monitoring the brightness of 150,000 stars selected from the 1.5 million or so in the catalogue. The Kepler field covers an area about four times the size of the full Moon. The fact that such an apparently small field has so many stars accessible to even a small telescope is a reminder that even smaller instruments can contribute much to our exploration of the vast unknown.

The spacecraft's unblinking gaze was combined with a camera capable of measuring the brightness of stars very precisely, making *Kepler* the most powerful planet hunter yet built. The discovery of planets around other stars has been one of the great scientific stories of the last few decades, with progress since the first unambiguous discovery in 1995 being both dramatic and exciting.* *Kepler* alone has been responsible for the discovery of thousands of likely worlds.

Most known planets, and all of the *Kepler* ones, are impossible to image directly. The planets are bright enough and close enough to us to be seen, but the dazzling light from their neighbouring stars outshines them. Attempting to pick out a planet close to a star which is thousands of times brighter is a nearly impossible task, like trying to spot the faint light of a glow-worm hovering next to a stadium floodlight. Instead, planet hunters rely on a variety of indirect methods to track down their quarry.

*Kepler*, for example, looks for faint blinks that represent the passage of a planet in front of its parent star. Such an alignment causes a brief drop in the star's brightness as seen from Earth, but such transits are not obvious. Even a large planet close to its star will still cover only a tiny fraction of the stellar disc, causing a dip of much less than 1 per cent in the star's brightness. The subtlety of the effect means that it is easier to get a reliable sighting without dealing with the distorting effect of Earth's atmosphere, which is why *Kepler* is a telescope in space. Even in such ideal circumstances, though, the odds of seeing a transit are low. Most possible orbits won't happen to take a planet in front of its

* Picking 1995 as the date of the first discovery is slightly controversial; there were claimed detections of planets before then, some in systems which did indeed turn out to be real, whether or not the original claim had enough statistical weight to hold up. There's also the strange tale of planets around pulsars, but whatever they are they're not normal and we can ignore them for now.

parent star as seen from Earth, and so projects like *Kepler* monitor many stars in order to increase the odds.

Once data is on hand, recording the brightness of many stars over an extended period of time, the task becomes that of finding any regular sequence of dips that might indicate the presence of a planet. Looking for a regular pattern is the kind of thing computers were built for and so this is an easy problem—or it would be if the data was clean. Instead, though, there are plenty of sources of confusion. The measurement is difficult and subtle, and noise in the camera can easily cause random fluctuations in signal big enough to mimic or to mask the signal. Worse, many stars vary in brightness. The well-behaved ones do so following regular cycles, but plenty are irregular and many have starspots, just as the Sun has sunspots.

Studying these changes in brightness is a science in itself, and can tell us a lot about the structure and evolution of stars, but as far as planet hunting goes they are a nuisance. The usual remedy is just to wait; once many separate transits have been seen at regular intervals, each one increases confidence in the reality of a planet's existence, but even then things are not straightforward and time is not always on our side. Planets in orbits close to their stars whizz around, racking up transits, but for most transits are separated by months or even years. Faster, better searches are needed.

In 2010 we were approached by Debra Fischer at Yale to see if we could help. Kevin Schawinski had moved from Oxford to a position there, and had been singing the praises of citizen science. Debra has a formidable record as one of the leading observers in the nascent field of exoplanet science, and explained at our first meeting that even the *Kepler* team themselves were reduced to inspecting possible planets by eye. (This is mentioned in the *Kepler* papers, though you have to dig through them carefully to spot it.) Rather than rely on a single expert classifying a small number of candidates that had been first filtered by algorithm, Debra and her colleagues were convinced it made sense to have a large number of volunteers sort through the whole *Kepler* data set.

This seemed mad to me. Only a few years into my Zooniverse adventure, I didn't have the confidence that I do today that people were donating their time to Galaxy Zoo solely because of a love of science. I still suspected that whatever the surveys said, and however grotty the images of the average galaxy were, people were still attracted by the experience of looking at images. The idea was that people would sort through not images, but graphs of brightness against time that astronomers call light curves. The proposition that people would be willing to give their spare time to look at graphs for fun, searching for something as indistinct and ill-defined as a dip in brightness, seemed like a stretch to say the least. The fact that we could add simulated planets to the data helped, as it at least gave people something to see as well as guaranteeing we could measure how effective the search was, but I was worried.

As a result, I spent days wondering whether we should be attacking this problem at all. I supposed eventually that even if we didn't find anything we could write a paper describing how good the *Kepler* team were at spotting planets. By this point, though, you should know what happens when I think a project will fail. Planet Hunters, the project we ended up building in response to Debra's challenge, was consistently our most popular project for years after it launched. The results were better than I could possibly have imagined—volunteers found nearly a hundred planet candidates that had been missed by the main search.

I'm following *Kepler* parlance: a planet candidate is defined as something that we estimate has a 95 per cent chance of being real. There remains a possibility that one in twenty of them will turn out to be contaminated by other effects, most likely by the presence of a distant pair of stars which eclipse each other and which happen to lie behind the star we think we're studying. The vast majority of our planet candidates will be real, though, and some are truly special. I have a soft spot for the world known by the ungainly moniker Planet Hunters 2b (PH2b), a Neptune-sized world which lies at the right distance from its Sun-like star such that any large moons would have a lovely, temperate climate, and would provide suitable homes for life (Figure 26).

One of the discoverers of PH2b was Roy Jackson, a 71-year-old retired police inspector from Gateshead. Asked on local television

**Figure 26** Planet Hunters 2b, a Neptune-sized world in the habitable zone of its parent star, as it might appear from an Earth-sized moon.

why he took up the unusual hobby of planet hunting, Roy's reaction was textbook British understatement. 'There's nothing to watch on TV', he said, 'and there's only so much gardening you can do.' If you want a sign of the speed of scientific progress, I think going in less than two decades from finding the first planet around another star to it becoming something you do from home while bored on a rainy Sunday afternoon is a pretty good one.

Even better, in this project too the unusual and the unexpected showed themselves. Planet Hunters 1b is a remarkable system with a terrible name, the only planet known in a four-star system. Two pairs of stars, each locked in a tight orbit around its partner, circle their common centre of mass. Planet Hunters volunteers identified a world which circles two of them. A few such circumbinary planets are known, but this is the only one in such a complicated system. As a result, just the fact that it exists tells us something new about how planets form.

That brings me to KIC8462852, the most unusual star in the Kepler database. Shortly after the space telescope began monitoring it, the star blinked. The drop in brightness amounted to almost 1 per cent of the star's normal luminosity; large for a planet, but not completely unprecedented. A few months later, a nearly identical dip was recorded. Three is normally enough to announce a discovery, and so a third dip would have provided evidence that something—either a large planet or a small and hitherto unsuspected star—was in orbit around the primary.

Instead, nothing more was seen for over a year. The star continued to shine brightly, as if nothing had happened. Either each of the two dips had been caused by separate objects, neither of which had yet completed a full orbit and returned to transit again, or the star itself was misbehaving. Among all the excitement of finding real planets, no one paid much attention to a star with a couple of glitches in its past.

Then the star dimmed dramatically. For a few short hours, it was suddenly 20 per cent fainter than before. It then returned to its former brightness, where it remained, as if embarrassed by its brief glitch, for more than a year. At this point all hell broke loose. The star dimmed, recovered, and then faded again. It looked like all was returning to normal, but then the star's brightness suddenly plunged again, once more fading by a factor of about 20 per cent (Figure 27).

This sort of thing kept happening, but eventually, after a few weeks of such baffling behaviour, the star returned to normal. Shortly after, the main *Kepler* mission ended. Not by design, but because the spacecraft's reaction wheels, responsible for keeping it pointing at the same patch of sky, had worn out. Monitoring of the star thus stopped, but the data safely received on the ground had brought it to the attention of Planet Hunters volunteers, who quickly realized they were dealing with something extraordinary.

Even inexperienced volunteers who ran across KIC8462852 realized they were looking at something odd. The time taken for whatever it was that was getting in the way to pass in front of the star was longer than you'd expect from a planet, and the dip and return weren't nicely symmetrical in the way they would be if

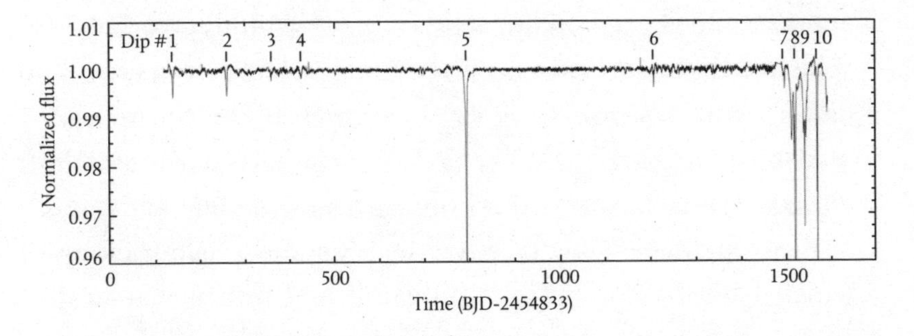


**Figure 27** The 'light curve' for the WTF star as seen by *Kepler*, showing the brightness of the object over nearly three years. The dramatic dips in the centre and near the end are almost unprecedented.

caused by a nice, round, boring planet. The task of investigating what was going on was taken up by a small group of volunteers, led by Daryll LaCourse.

Over the course of the project, this bunch of citizen scientists had learned to use many of the tools provided for professional astronomers to work with data from *Kepler*. One of the first comments on KIC8462852—now renamed the WTF star*—pointed out the similarity to a set of unusual objects called 'heartbeat stars' that the group had already investigated.

These are bizarre systems in their own right, with a name deriving from the fact that the graph showing their brightness over time—what astronomers call a light curve—resembles the trace of an ECG in a casualty ward. Periods of stasis are broken by a sudden increase in brightness, then a decline, and then a return to the long-term norm. Planet Hunters volunteers learned about these stars when watching Jim Fuller and Dong Lai, researchers from Caltech and members of the *Kepler* team, present at a conference in California. The meeting was streamed live on the internet, and so the audience involved volunteers scattered around the globe as well as the gathered community of planet-hunting professionals. Before the end of the talk, which described the first two heartbeat stars, the volunteers realized they'd seen similar behaviour and started pulling a list of these oddities together.

The heartbeat stars turn out to be unusual binary stars, with a smaller secondary star on an elliptical orbit. When the secondary swings close to the primary, the latter rings like a bell, and the resulting changes in brightness appear as the distinctive pulses we see in the data. This happens on a regular schedule, as the

* Following negotiation with a journal editor who insisted that it was policy that all acronyms be spelt out, it was agreed that WTF stands for 'Where's the Flux?', neatly referring to the central mystery presented by the star.

ringing repeats with each swing of the smaller star past the larger. This is interesting, and such behavior valuable in trying to understand stellar interiors, but something more complicated is going on with the WTF star.

Daryll found a clue in the data that was available on the web. Most of the objects in Kepler's target list had been extensively studied in preparation for the mission. In many cases, the properties of any planets found can only be pinned down if the stars themselves are understood, and effort has already gone into excluding stars whose inherent variability would have hidden likely planets. As a result of all this work, Daryll could tell that this star was brighter in the infrared than stars of its type usually are.

Unexpected brightness in the infrared usually means that there is a disc of dust, leftover from planet formation, in orbit around a star. As the dust absorbs light from the star, it reradiates mostly infrared radiation; hiding a star behind dust therefore usually results in the system appearing dimmer than normal when viewed in visible light, but brighter in the infrared. This is one reason that astronomers studying star formation normally turn to longer wavelengths, hoping to be able to observe stars still embedded in their embryonic cocoons. The infrared excess suggested that the neighbourhood of the WTF star was a dusty place, and Daryll realized that this might be the key to explaining its bizarre behaviour.

He suggested that there really was a planet in orbit around the star, but that the planet was itself surrounded by a dust disc. That seems sensible enough. Just as planets form from the disc of leftover dust and gas which surrounds newborn stars, so a newly formed planet might be surrounded by a disc of leftover material from which moons might form. Our own Moon probably had a more violent origin, coalescing from the debris of a collision between the proto-Earth and a Mars-sized object, and the two moons of Mars seem to be no more than captured asteroids. Large planets such as Jupiter seem to have formed their systems of large moons more directly, though, and a large planet with a dense disc of material passing in front of a star would certainly block plenty of the star's light. If the geometry of the passage changed each time, you might be able to explain the observed differences each time the WTF star dimmed or flickered.

The attractive thing about this proposition is that it could explain an almost arbitrary pattern. The disc might have a gap at its centre, between the planet and its inner edge, just as there's a gap between Saturn's rings and the planet itself. The team flying the *Cassini* spacecraft, which took several plunges between rings and planet at the end of its life, called it the 'Big Empty', so it should be no surprise that light would shine through such a gap, adding to the complexity of the behaviour during an observed 'blink'. Saturn's rings also have gaps within them, shaped by interactions between ring particles and the myriad tiny moons which surround and shepherd them. Add the same sort of thing to the WTF system, and you might have a chance of explaining what's going on.

By the time speculation had reached this state, with Daryll and others drawing possible models for the rings, the Planet Hunters team themselves became involved, most notably Tabby Boyajian, then a Yale postdoc, who led the professional end of the effort to solve the mystery of this most unusual star.* The dust disc explanation felt wrong from the start; every piece of information we had in the Kepler Input Catalogue pointed to KIC8462852

* Because of Tabby's efforts in leading the work on the star, it's sometimes known not as the WTF star but as Boyajian's star, which I rather like. 'Tabby's star' also gained currency, but I like the authoritative and official sound of using her surname.

being a perfectly ordinary, stable, and middle-aged star, while dust discs are almost exclusively the property of younger objects. Worse than that, the fact that two of the dips accounted for almost 20 per cent of the star's light meant that the obscuring object had to be enormous. If enough dust existed in a disc to create such a large dip, it should be stonkingly bright in the infrared—and it wasn't.

So there's no dust disc. And the star appears to be perfectly normal, with nothing in its colour or spectrum marking it out as one especially likely to behave oddly. The team led by Tabby checked that there were no signs of camera malfunction. Neighbouring stars appeared to maintain a nice, constant brightness, and when, driven by some magic combination of desperation and paranoia, they checked which pixel each observation of our star landed on there was no obvious pattern that might explain the observed dips. In the paper we put together announcing the discovery, and on which seven separate citizen scientists appear as authors, we fairly reluctantly nailed our colours to a hypothesis that suggested that the dips we observed were the result of a string of comets.

Comets have a lot to commend them. For starters, they're less bright in the infrared than one would expect a dust disc to be, and that means you can hide enough stuff to cause big dips in brightness without exceeding the infrared limit set by the observations. Our comet would need to be broken into bits, so that each piece could be responsible for an individual dip, but that's ok. Breaking up is something that comets tend to do. Comet Schwassmann–Wachmann 3, for example, survived for sixty-five years after being discovered by two German observers in 1930, but broke into four pieces in 1995. By 2006 it was in eight separate pieces and seems to be in the process of crumbling entirely. Comet Biela, a spectacular sight in the nineteenth century, split somewhere around the middle of the century and had disappeared completely by the time of its predicted return in 1859. Both Schwassmann–Wachmann 3 and Biela even produced short-lived meteor showers, their remnants burning up in the atmosphere as Earth crossed their orbits.

The most famous of comet breakups was that of Shoemaker–Levy 9 (SL9), which came too close to Jupiter in the early 1990s. By the time it was discovered, the giant planet's gravity had split it into a string of separate nuclei, each on a collision course with Jupiter itself. The impacts happened on the far side of the planet as seen from Earth, but I will never forget the experience of turning my small backyard telescope to the planet a few hours later and seeing clearly the striking bruise left in Jupiter's atmosphere by the impact of the first large piece of the comet. I dragged my parents out of bed so they could take me to the larger telescope at school, and marvelled at a sight not seen for centuries.

More recent amateur observations have established that asteroids and comets hit Jupiter at least a couple of times a year, but SL9 was special because of the size of the comet and the sheer drama of the event. For a week or so, impact after impact caused bruise after bruise in the giant planet's atmosphere, many of which remained visible for months following the impact.

These experiences made it seem sensible to us that a comet might have happened to break up just as *Kepler* started observing this particular patch of sky. People who actually understand comets disagreed. A typical comet nucleus is a small thing. That visited by the European Space Agency's *Rosetta* probe and its famous bouncing lander, *Philae*, is just a few kilometres across.* Our comet would have to be the size of Ceres, the largest body in

* Churyumov–Gerasimenko, since you ask, but commonly known as Chewy-Gooey until it was pointed out that Churyumov and Gerasimenko, its discoverers, might not be amused.

the asteroid belt. When discovered, Ceres was large enough to be considered a planet, but the increasing flood of discoveries quickly relegated it to being just an asteroid, a nineteenth-century parallel to the plight of Pluto in the twenty-first. Both are now technically classified as dwarf planets (much to the chagrin of a loud and very vocal minority of the planetary science community and the wider cacophony of shouty people online).

So we had either found the largest comet known, and done so just as it started to disintegrate, or we had no idea any more what the WTF star was up to. Others had ideas, and we heard a lot about one of them in particular. Jason Wright from Penn State and his colleagues thought that our discovery fitted perfectly with a research programme they had underway, and the title of their paper was certainly eye-catching. It's called 'The Ĝ search for extraterrestrial civilizations with large energy supplies IV. The signatures and information content of transiting megastructures'. It's that last word that does it; mention finding alien 'megastructures' and the world and its dog starts to pay attention.

Specifically, the word 'megastructures' turned out to be catnip for journalists. It sounds just technical enough to make the story appropriately sciency, while not being so technical that it puts people off. The paper Jason and friends published (in the *Astrophysical Journal* no less—the premier US venue for astronomical research) spends most of its time talking about how one might, if so inclined, use data to distinguish a transiting alien space station from the signature of an ordinary planet. The logic is that any sufficiently advanced alien civilization would want to make use of as much energy as possible, so rather than idling away on the surface of a planet like ours would seek to surround their star with fleets of orbiting solar panels. Often called in science fiction a Dyson sphere, a spherical shell surrounding a star would be unstable. It's best to think of many individual orbiting

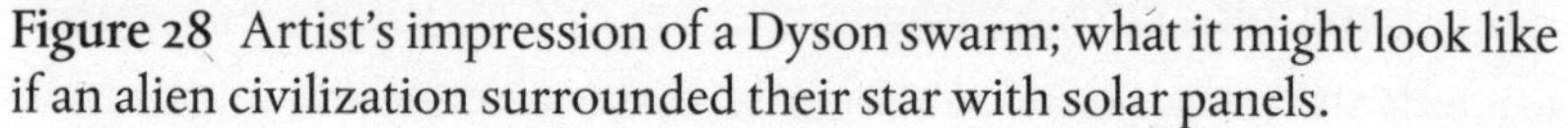

**Figure 28** Artist's impression of a Dyson swarm; what it might look like if an alien civilization surrounded their star with solar panels.

spacecraft arrayed into a much larger 'megastructure'—what the physicist Freeman Dyson called a swarm (Figure 28). Either way it would be a spectacular feat of engineering on the grandest of scales, but as the authors of the paper pointed out, clusters of swarming alien spacecraft would make a pretty good explanation for exactly what we see in the WTF star's blinking.

So Planet Hunters volunteers may have been responsible for the discovery of alien intelligence, the most significant moment in astronomical history. Have they really? The press wrote the story up as if astronomers had seen green tentacles waving back from a passing spaceship. That fuss made the star famous, and ultimately led to an appeal on the web to fund Jason and Tabby's efforts to keep an eye on their new favourite star.

Screaming 'Aliens!'—or in this case, having the press scream 'ALIENS!' on your behalf—turns out to be a good way not only to attract those who might want to donate to your research, but also to get other astronomers to notice what you're doing. There can't have been a department anywhere in the world that didn't discuss the star, if only in idle chat by the coffee machine, but conversation led to action at Harvard, where the observatory keeps a stack of historical images of the night sky. Most exist in the form of photographic plates, enormous things that could be strapped to the end of a telescope and exposed to record dim starlight. The Harvard observatory has spent a lot of time and energy scanning these things, turning relics sitting in an archive into useful, digital data, and it was quickly realized that the WTF star appears in more than ten historical plates, dating back to the late nineteenth century and stretching forwards to 1970 or so.

These historical records revealed the startling fact that the star has been gradually fading over the course of the century. This result started an enormous row among the handful of experts on such data, who disagreed about how long-term storage and the process of digitizing the plates might have affected the results, but more recent, careful analysis of the *Kepler* data seems to confirm the observed trend. The star is fading slowly and seemingly inexorably, regardless of the dramatic sudden dips that had drawn attention. My scientific instinct tells me that we're looking for an explanation that ties together both unusual behaviours—slow fade and sudden dips. Having one star behave oddly for two different reasons seems like a stretch, and so I reckon we're searching for just one answer.

Clearly the slow fading of the star has implications for any alien civilization too. Perhaps they are still constructing their star-circling space station. In a note we published in the *Journal of Brief Ideas,** with tongues firmly in cheek, Brooke Simmons and

* This is a real thing—you can find our paper here: <http://beta.briefideas.org/ideas/424bb64cf38eb9d7db0dae57dec3d28d>.

I calculate their progress, assuming that the end goal is a full Dyson sphere which completely captures the star's light. Assuming construction doesn't slow down (or speed up), they've got about 700 years left. We also noted in the paper that this probably meant that elections on any worlds responsible probably occur less than once a millennium, it being hard to fund infrastructure projects anywhere if they last longer than a single electoral cycle.

By the end of 2017, though, there was still no clear consensus as to what was going on. New infrared observations suggest a surrounding dust cloud might be responsible for the slow fade, but not the dips. The leading hypothesis in my mind is that the star has recently swallowed a planet. Such an event, as modelled by exoplanet astronomers whose imagination knows no bounds, would apparently cause the star to brighten and then to slowly fade. Any remaining rubble, left over from the inevitable disintegration of the planet, could be responsible for dramatic dips—the explanation accounts for both halves of the puzzle, but more evidence is clearly needed.

Specifically, we need data taken during one of the dips by telescopes larger than *Kepler*. A worldwide network of robotic telescopes has been employed to keep an eye on the star, and plenty of other professional and amateur observers have joined in too. For a couple of years, nothing happened. And then, one otherwise unremarkable day in May 2017, the star dipped once more.

This threw Tabby and her colleagues into a frenzy of activity. Just as we'd relied on the black market in telescope time to get that initial spectrum for the Voorwerp, so the team started calling, begging, and pleading for people to observe the star. Some of this activity happened quietly, as applications for what's known as 'Director's Discretionary Time' (available slots in the personal gift of the observatory director) and programmes which allow one to observe 'Targets of Opportunity' went in, but it also consisted of frantic Twitter activity, with Tabby and others posting the latest data that showed the dip in progress and asking for help.

That meant we—the world—watched as the star dipped, recovered, and then dipped again. Debates broke out about whether the dip was the same shape as on previous occasions. Spectra were obtained and slowly the star returned to its normal brightness. What do the results tell us? Well, the mystery remains, but we know one thing—there is no alien megastructure orbiting this particular star. We know that because observations were obtained during these 2017 dips in brightness which showed that how much the star dims depends on what colour filter you observe through. If you can detect only red light, you'll see the star dim less than if you can only detect blue. This is not behaviour readily caused by solid objects (although some have—I think in jest—suggested that that we're seeing alien Christmas lights hung on the 'outside' of their space station). Rather, it likely indicates that the light is being blocked by something like a cloud of dust. We don't have all the details yet, but we seem to be beginning to close in on the end of the mystery.

Stories like those of the Voorwerp and WTF star are fascinating to me because they are the kind of thing I imagined astronomers did when I was that small kid with a telescope. The discoveries may be more about finding new ways to investigate the Universe's mysteries rather than something to name after me, but who cares? This image, and this way of working, couldn't be further from the usual rhetoric about the onward progress of 'big data', in which we solve problems by writing database queries.

So what had to happen for these discoveries to take place? First, large surveys—the biggest of big data—need to exist. You can only find the unusual by fishing in a very large sea; fewer than 1 per cent of galaxies have a Voorwerp and the WTF star is at the very least only one in 150,000, and so we do need to collect many, many data points. Then we need to pay individual attention to each, asking if it is unusual and worthy of interest.

This step is what the Zooniverse projects provide. Professional astronomers couldn't possibly look at each system individually, and—even better—people like Hanny become advocates for their objects. It's not that the public are more curious or interested than professional scientists, but my guess is that citizen scientists are more likely to interrupt their core task to consider a curiosity. We're all walking along the same path, if you will, but if you're not being paid by the mile you're much more likely to stop and smell the flowers, and notice an interesting insect or two while you do so.

Enough data, and enough citizen scientists, and you can spot the truly interesting stuff, but that's not sufficient. Not everything that is interesting is significant. Galaxy Zoo volunteers found galaxies shaped like letters of the alphabet, and Alice Sheppard, our moderator, adopted a beautifully penguin-shaped galaxy as her avatar. Plenty of volunteers have been taken aback by the sight of a bright green streak shooting from one side of an image to the other—not (sadly?) an alien space laser but the track of a satellite captured by the telescope by accident. Here, the forums that are attached to the project—and more to the point the communities that gather in and around them—are extremely important.

On projects where we've seen this sort of serendipitous discovery happen, it's usually been because of a community of citizen scientists who can sort through the novel discoveries of thousands of classifiers, distinguishing the humdrum satellites from the unusual galaxies. Often, this group do plenty of work before turning to the professionals, aided by access to raw data from public surveys; Daryll's investigation of the strange WTF star is a case in point. Together with their classifying colleagues, citizen science communities provide a wonderful filter to identify the most unusual and interesting objects. That's what I think is the greatest single reason for trying to preserve citizen science like the Zooniverse as machines get better; collectively we can find not only unusual objects but also new questions.

One way of thinking about this was given to me while listening to talks by the team behind another Zooniverse project, in an area far from astronomy. Shakespeare's World asks volunteers to transcribe material from the sixteenth-century collections held by the Folger Library in Washington, DC, in order to try to help understand what life was like back then and to trace the history of language. The project is a distant descendent of Old Weather, and, as in that project, participants have paused along the way to investigate all sorts of curious finds.

A volunteer whose screen name is mutabilitie (sadly, we don't know their real name) found a 1567 letter containing the lines 'Albeit I do assure you he is vnsusspected of | any vntruithe or oder notable cryme (excepte a whyte lye)', the oldest recorded instance of the phrase 'white lie' by more than 150 years, now recorded in the *Oxford English Dictionary*. You can also see why the appearance of a recipe 'To make mackroones or portugall farts' drew the attention of volunteers; 'farts' were little, light pastries, no more than puffs of air.

One of the researchers concentrating on recipes for farts and other things, Lisa Wright, explained to me that what the volunteers were doing was a technique known as 'close reading', common enough in studies of literature. The idea is to pay attention to each individual word in a text, working out what each contributes as well as considering it in its own right. Of course, the problem is choosing where to focus, but here the volunteers were providing close reading at scale. Because we have a crowd of volunteers, we can scan every word in a large corpus of material, or pay individual attention to a million galaxies. Unlike most attempts to use large data sets to do research, the point isn't to take the traditional close reading—close study, if you prefer—of individual words or objects and replace it with some database query, clever visualization, or statistical analysis, but rather to keep the traditional method of analysis alive. The way the crowd behaves allows us quickly and collectively to home in on the examples where it will be of most use.

Citizen science, seen like this, is a way of finding the interesting stuff and focusing rigorous, sustained, detailed attention on it. By being distracted, we can appreciate and try to understand the unusual. It's a wonderful way to make discoveries, and a lovely way to do science. But does it have a future?

# 9

# THREE PATHS

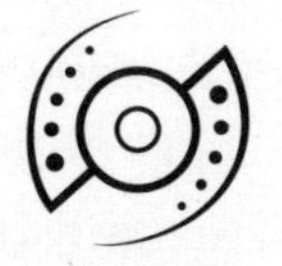

The Zooniverse project has grown so fast and so far that a decent description of all of our projects we've done would have filled this book and more. I recently spent a day trying to complete a single classification on the more than seventy projects currentlylive and wound up exhausted and overstimulated. In that day, I'd helped biologists map the ocean, had transcribed ancient Hebrew texts, done all manner of astrophysical tasks, and measured stuffed birds from the Natural History Museum collection. I used to know every project intimately, spending time thinking about the design and data of each, but especially since we launched a tool that allows researchers to quickly build their own projects instead of relying on web developers, those days are long gone.

What people are using our tools for is constantly humbling. Brooke Simmons, the member of the Galaxy Zoo team who led the work on bulgeless galaxies described in Chapter 4 and now an astrophysics lecturer at the University of Lancaster, has led an effort to try to build what she calls the Planetary Response Network. When a natural disaster happens, Brooke works with networks of first responders who will be flying in to help with the relief efforts. Sometimes, as with the earthquakes in Nepal in

2016, the area affected is so remote that there are literally no maps. In other cases, like during the Caribbean storms of 2017, the maps need rapid updating to reflect the effect of the disaster on roads, buildings, and even people.

It's amazing stuff, made possible by the profusion of Earth-observation satellites that now exist, capable of imaging any part of the globe at short notice. The companies that run these constellations of cameras—most notably the Californian start-up Planet, who have more than 150 orbiting imagers—are also pretty generous with their data, making them available for genuine humanitarian efforts. The results are impressive too; in the case of the Nepalese earthquake Zooniverse volunteers identified an otherwise unmapped village within the affected zone, directing personnel from the UK charity Rescue Global to a place they might not otherwise have visited. The results of the Caribbean deployment were less spectacular, but those going to the aid of storm victims in Guadeloupe, Dominica, and Puerto Rico used maps which included contributions from volunteers. We're now working hard to make it easier to include new sources of images so that we can respond faster in the event of future crises.

When Galaxy Zoo started, we couldn't imagine doing anything like this. While Earth observation—taking pictures from space of our home planet—has long been a major reason to launch satellites into space, the availability of images that are sharp enough to pick out details such as a landslide blocking a road was until recently almost exclusively limited to the military, along with other government agencies and their partners. Even when high-resolution images were released, they were typically out of date; scheduling a sudden imaging campaign following a disaster was next to impossible. Now, the situation is completely different. I'm sure military technology has moved on too, but the small-satellite revolution has changed the game entirely.

The way that space technology has moved from being about cutting-edge, specialist tech to being about clever reuse of components developed for other things—including your mobile phone—is a story whose consequences I think we're still trying to understand, but its effects are becoming clear. Because satellites are cheaper, more can be launched.* Because more are up there, the chances of one flying over any given location in the next hour or so have increased dramatically, so up-to-date images are easier than ever to obtain. At the moment, most of this data is private, used by commercial companies for everything from assessing traffic flows to directing fertilizer spraying in fields of crops, but I think the day is coming when either public space agencies like the European Space Agency or NASA, or private companies with different business models, will make large amounts of high-resolution Earth imagery open to anyone for free. At that point, as long as the tools to use this wealth of data are also made available, we should expect a flood of citizen science projects similar to that seen in astrophysics in the last decade.

What might those projects look like? There are already examples of craters associated with asteroid impacts being spotted in satellite images; one, the Kamil Crater in the middle of the Egyptian Sahara, is forty-five metres across and sixteen metres deep, yet was first identified by scientists using Google Earth! We have had projects pitched to the Zooniverse that want to use satellite imagery to assess the number of street traders in southern African cities, and thus work out how much of the country's economic activity might be taking place in this informal way rather

* The increasing number of ways of getting your satellite to space helps too; Elon Musk's SpaceX have played a large role in making it cheaper to launch things, but there are plenty of innovative companies building small and medium-size rockets capable of launching whole constellations of satellites.

than in the more traditional economy that shows up in tax returns. The number of suggested projects which involve assessing human activity around the world is increasing, though so far we haven't promoted any of these projects on the Zooniverse platform; we need more advice from people who aren't moonlighting astronomers before I'm comfortable.

More obscure projects are possible too. My favourite example is an attempt to settle what is apparently a vigorous scientific dispute between two sets of researchers. One group believes that cows are sensitive to magnetic fields, and will tend to align with the prevailing field (the magnetic field, not the farmer's field), while another thinks this is nonsense. The only way to settle the matter is to collect more data. Cows are visible in many satellite images, so all one would have to do is find a set of volunteers willing to mark bovine orientation on a sufficiently large number of images.

The examples and possible projects seem endless. But things are very different now from the time the astronomical citizen science revolution was beginning, as you'll have gathered. Ten years on, the idea of it being a novel thing to invite the public to participate seems quaint, and new projects don't get the kind of media attention that drove the initial success of Galaxy Zoo. More to the point, machine learning has improved to the point that building an image analysis project without considering the complex interactions between humans and machines seems negligent, if not downright unethical, in the way it would waste people's time.

So what future does citizen science of the sort carried out by the Zooniverse volunteers really have? I'm going to concentrate on my home turf, on astronomy, and even here I think there are three different possibilities, three possible paths that we might find ourselves on which lead forward from where we are now.

Which one we end up with will depend on how much effort we're prepared to put into open collaboration, and what kind of science we want to do. I hope that as many projects as possible will make the right choices; we certainly intend to.

The first scenario is the one we will reach if we don't do anything about it. It seems obvious that the current improvement in machine learning, powered by research carried out not only in the computer science departments of universities but also by the increasingly large machine learning teams at Google, Facebook, and in the rest of Silicon Valley, will continue. Companies from these giant firms all the way down to the newest start-up clearly see being ahead in artificial intelligence as essential for twenty-first-century business, and while at present that mostly means having a larger labelled training set to teach your robot new tricks, it also includes innovation in techniques, many of which are aimed at the kind of problems we encounter in astronomy.

You can make a reasonable case that machines—specifically convolutional neural networks—can now be trained to do basic galaxy classification as well, if not better, than the crowd. If you just want to split spirals from ellipticals, for the vast majority of systems no human intervention is necessary, and it is beginning to look likely that a system trained on one survey, such as Sloan, may be able to cope easily with galaxy images coming from completely different surveys and therefore with different depths, colours, and characteristics. This is happening partly because splitting spirals from ellipticals is the easiest of the problems that Galaxy Zoo posed to its crowd of volunteers, and partly because it is the question for which we have the largest volume of data with which to train the machines.

It shouldn't be a surprise that this question might pass from the realm where we need human intervention to that where machines rule. The same thing happened to the task of separating images of galaxies from stars, the latter appearing as sharp points, easily contrasted with the fuzzy blobs of distant star systems twenty or thirty years ago. I wrote in Chapter 6 about the supernova-hunting project that put itself out of business in just this way. What of questions where the training data is not so abundant?

My group in Oxford now includes a PhD student who is an expert in machine learning (and critically, articulate in explaining it to the rest of us). Mike Walmsley has just finished working on a specialized neural network that can find the faint structures around galaxies which indicate a past merger. Looking for these faint tails of stars is important if we want to understand how normal galaxies react to collisions—it's sort of the opposite technique to finding the bulgeless galaxies I discussed in Chapter 4. Bigger collisions (those with more massive galaxies) leave more debris, so at least in theory there's also the chance of reconstructing the crash that led to stars being scattered out of the main galaxy itself, if only we can find them. The trouble is there are very few surveys where experts have done the painstaking work of sorting through the images themselves.

Nonetheless the results, despite the handicap of a small training set, are pretty good. The network is indeed capable of finding galaxies which show signs of a merger. It's not perfect, matching expert classifications 80 per cent of the time, but that's a huge advance on where we were before. In the old days of 2007 or so, we'd have set up a citizen science project to gather more training data and to try and improve this figure. A few years ago we might have looked at how to combine human and machine classification, like we did in the supernova project. But in this machine-optimistic scenario, another year or so's work will break the back of the problem, and we can expect the robots to win before too long. If neural networks really can be adapted to deal with such small training sets, then we won't need large numbers of classifications from volunteers.

Progress might come from more of Mike's work, which uses a new kind of neural network introduced to us by a colleague in computer science, Yarin Gal. This network not only classifies things, but can tell us how certain it is about its classifications. It's thus producing data which is of the same kind as that produced, collectively, by Galaxy Zoo volunteers. By the time you read this, we'll be running it alongside the main project, and incorporating its results into our decisions about galaxies.

Another major area of research in machine learning is in finding unusual objects. Actually, that's not quite true. Finding unusual objects—the images in the original Galaxy Zoo data set of nearly a million galaxies that look least like the others, for example—is not a hugely difficult problem. As I wrote earlier, the difficult bit is finding unusual objects which are actually interesting. It's one thing to pick out the images where the camera malfunctioned, where a bright star overwhelmed the chip or where someone turned a light on by mistake, but quite another to find the peas and the Voorwerp among that pile of images which are occasionally visually interesting but mostly scientifically junk.

Still, progress is being made. Techniques which use 'clustering'—sorting similar images into piles—look promising. If you end up with many piles with a few images in, it's not a huge amount of effort to decide which of these outliers are truly interesting. Future surveys might do this as a matter of course, with their professional astronomers presented with a few representative objects from each class for consideration.

Perhaps this focus on the unusual is in any case wrong-headed. If astrophysics is heading for a future where we produce truly enormous data sets then we might have no choice but , like Dr Strangelove, to stop worrying and learn to love the algorithm.

Maybe we can get more insight from things that occur often than from the odd weird exception. Particle physicists at the LHC are, for the most part, already living in this future; as mentioned in Chapter 1, if some completely unexpected cascade of particles happens in this most massive and sophisticated of experiments, it will be discarded by a system looking for specific triggers. The LHC detectors simply couldn't operate any other way without being completely overwhelmed by noise.

Cosmologists, too, seeking to discover type 1a supernovae so as to measure the effect of dark energy in the acceleration of the Universe's expansion (see Chapter 6) may not mind if explosions that don't fit the expected pattern are discarded. If you can find enough supernovae of the right type, you may even get better results by assembling a nice, well-behaved group rather than including anything odd. For predictable science, where we're testing well-defined hypotheses—something that would fit well into the science fair I described in Chapter 1—trusting the machines and hoping we end up in this future might well be a sensible way to go.

A second possible future is one in which, though machine learning continues to improve, we never really break free from the tyranny of the training set. The techniques that are driving the artificial intelligence revolution simply are, like an easily distracted student, dependent on being walked through example after example after example.

There are some ways of dealing with this. Techniques like transfer learning, where a neural network or other solution is trained on one survey before most of its guts are used to construct a new network capable of dealing with a different data set, do make things easier. A network trained to recognize animals in the Serengeti will do pretty well when deployed on images of wildlife in the US; though the species are different, the layers of the network that identify the animal amid the background will be shared between the two problems.*

For a project like the LSST survey, where there are a thousand different scientific investigations that all need access to the same, consistent data set and where rare objects matter, it's less clear what the solution is. After all, finding unusual and unexpected objects is part of the reason we build telescopes like this; whenever we've done something fundamentally new, in this case monitoring such a large area of sky this frequently with such a powerful instrument, we have found new things.

And if LSST is going to challenge machine learning, then once the data from the radio astronomers' new toy, the SKA, starts flooding in then we'll really be in trouble. In this scenario, the problems faced (and caused) by scientists in general and by astronomers in particular are odd enough that whatever Silicon Valley gets up to we'll need help ourselves.† This means that well into the next decade, we'll need plenty of classifications from humans and their expert pattern recognition systems. Indeed, looking at what's coming, the existing effort across all Zooniverse projects won't be enough to cope.

We need to get smarter if, in this reality, we're going to preserve a space for citizen science. Probably the easiest way to do this is to recruit more volunteers to help. (Despite this being a vision of the future I'm making up, let's assume that even in this universe it's not the case that millions of people have read this far

* This sort of work is being led for the Zooniverse by Lucy Fortson's group at the University of Minnesota.

† This isn't completely unrealistic; there aren't too many cases where the most important things are the rarest objects, or where such precisely accurate classifications are required. If Facebook identifies the wrong friend in a photo, it's at worst slightly embarrassing, and is unlikely to lead you to predict the wrong future for the Universe.

so as to be inspired to rush to the keyboard and contribute). I'm sure there's more we could do,* but to really tackle the bulk of LSST let alone SKA data we'll need an enormous increase in the amount of effort available.

The answer may be staring us in the face. If human beings are game-playing creatures, then maybe we should build games rather than citizen science projects. Indeed, the first moves in this direction have already been made. Eyewire is a project run by researchers at MIT, who want volunteers to help map the complex structure of neurons in the brain. Volunteers see slices of the complex tangle of cells and are asked to separate the structures visible in the images from the background; additional help and complication is provided by the fact that these are in fact three-dimensional objects. It sounds complicated, but the team have provided an engaging and interesting interface that has attracted tens of thousands of volunteers to help, producing results that, in a preliminary study, were impressively accurate.

Eyewire participants also chat to each other, and to helpful chatbots which offer advice, in real time while they're classifying. It's a much less isolated experience than our Zooniverse projects, where the act of classification is performed in sacred solitude so as to prevent groupthink (as we've seen, discussion and collaboration through our forums happens after the initial classification is recorded). Eyewire volunteers also score points for their participation, and an ever-growing set of challenges and competitions aims to make the game more engaging, and to bring classifiers back for more.

A recent email newsletter sent to me and the worldwide network of my fellow Eyewire volunteers gives you the idea. During

* Have you considered buying a copy of this book for a friend? Or three? Or for everyone you know?

the summer of 2018, alongside the real thing in Russia there was an Eyewire World Cup. Participants representing a country had their effort counted towards their team's total, and could win 'buckets of points, six new badges and speciality swag [they'd] only be able to get if [they] participate'.

These are the techniques of modern software development and game design, being used here to drive people towards taking part in a scientific project. I'm an enthusiastic participant in the project, so please don't think that I consider the idea of point collecting and competitions beneath me. The reality is quite the opposite; their techniques work especially well on me!*

Others have gone further, and made a game of the science itself. Probably the best known of these projects is an old one, predating even Galaxy Zoo. Fold.it asked volunteers to investigate the three-dimensional structures of proteins. In many cases, we know the basic chemistry of these important biological molecules in the sense of being able to write down what connects to what. However, secondary effects as the atoms bond together will cause the protein to twist and buckle in a way that is currently very hard to predict; it's impossible to calculate, and any automated search for a likely solution runs the risk of getting stuck in a local minimum, a possible solution that looks plausible (technically, it's likely better than any solution that is similar to it) but which has not been tested sufficiently to find out whether it is overall the best.

Exploring a vast range of possibilities to find a good solution to a problem like this is another type of task that humans have evolved to be good at, just like the more basic pattern recognition

* I am, in fact, a sucker for this sort of thing. I have an enormous pile of coffee shop loyalty cards from places I will never again visit, and have used the Foursquare app to check in everywhere I've been since 2011.

that we in Zooniverse have been using all this time. Once a structure is proposed, it is easy to calculate its energy, based on the interactions between the various components. The game is to look for the lowest energy structure, as we trust nature to have found a way to fold proteins efficiently. All this effort is important because it is the three-dimensional shape of a protein that determines how it interacts with other molecules, particularly in the complex and not fully understood dance that is molecular biochemistry.

The results from Fold.it have been great, with players often outperforming the best computer science efforts at attacking the same problem in large competitions and challenges designed to test protein-folding methods. Sometimes the best players turn out to be those with some sort of relevant expertise, but more often the game finds people who turn out to have an instinct for how to play. Because the 'rules'—things like the angle at which hydrogen atoms can be placed—are encoded in the game itself, Fold.it players don't need to know any chemistry at all.

It's a neat solution, and the game is actually quite fun to play, even if I can't get past the first few levels. I've never been patient with puzzles, but it seems I'm not that typical. A few years ago, when I visited the Fold.it team at the University of Washington, they told me that at any one time a few people are deep enough into the game that they're providing real and useful results, while most players are still learning. If the number of useful players drops too far, the team will run competitions or advertise to encourage a new cohort of Fold.it players to work their way deeper into the system. The entire structure of the game is a conveyor belt designed to carry the best players onward to the point where they're working on scientifically useful data.

It would be possible to play Fold.it without realizing it had a scientific purpose at all, though I doubt anyone does so. Other teams have gone even further, disguising citizen science projects within existing games. Probably the most ambitious example is a Swiss project that created a mission within the science fiction-themed online multiplayer game, Eve Online. Players of the game can choose to review data from *Kepler* in the hope of finding a planet, but also in order to receive rewards in the form of the game's internal, online currency. The experience is noticeably a bit odd, but in essentials indistinguishable from the experience of completing one of the other missions within the game world itself.

With millions of people taking part in such games, here, perhaps, is the crowd we need in order to cope with the data sets of the future. In this imagined future, projects like those hosted on the Zooniverse will become both more ubiquitous and almost completely invisible. In fact, the more invisible they become the better, as the more seamlessly they can be integrated into the games we're playing anyway the more people will take part. Instead of having to make the choice to participate in science, something which many people find intimidating, it will just happen.

Will this work? Maybe. Half a million people took part in the Eve Online planet hunting experiment, though I haven't seen any discoveries come from it yet. That's not too surprising, as these things take time, but it will be the acid test of whether the project has succeeded. (A similar effort, which involved more than 300,000 players in the task of labelling features in high-resolution images of cells, has recently produced a paper which shows that the technique works, at least in this one case.) Even our modest experiments with gamification in the original Old Weather project (described in Chapter 4) seemed to work well. All we did was give people a rank when they started transcribing records from a ship, and yet it seems to have encouraged some people to work very hard indeed. One 'ship' in the project was, I'm pretty sure, a building—a training facility given, as is normal in naval tradition, a ship's name. Despite the fact that it didn't go anywhere, people dutifully worked their way through the log book. (I haven't followed this up, because the implications of being able to inspire people to work their way through the log of a building pretending to be a ship scare me a little.) With the help of games designers, maybe we can hide enough tasks that citizen science even at the scale needed for these big surveys will become possible, and all without anyone knowing they are participating.

This second future reality is efficient, and science gets done, but I'm not sure I like it. Actually, I'm certain that I don't. I've tried in these pages to convey not only my excitement about the results of our projects, but that of the people who get caught up in them. Participation in Zooniverse projects gets science done, but it's more than that. A study led by Karen Masters, Galaxy Zoo's project scientist (and now the elected spokesperson of the entire Sloan Digital Sky Survey), which asked questions of volunteers while they were classifying, showed that people who participate in such projects learn things.

They learn things, in fact, that they couldn't have learned from the projects themselves. In other words, taking part in, say, Galaxy Zoo, inspires people to go out and seek out more information about the Universe. The projects act as an engine of motivation, creating a cohort of people who are actively seeking information they never knew they wanted. Think of Planet Hunters finding the details of transiting planets, or Old Weather participants digging into naval history, or any of the other byways and distractions we've inspired. Gamifying the experience—hiding the science behind a thin veneer of play, making it feel less like real science and more like any other game on your phone—might make projects more efficient, but it would kill this most important side effect of participation stone dead.

You can see this in the studies researchers have done of the effect of even the simple gamification—the ranks on board the ship—that we added to the Old Weather project. Volunteers they interviewed said that the game worked as designed, in that it made them more likely to do more work, but there was another, more disturbing result. Instead of describing participating as fun and interesting, they suddenly used language that made it seem like work. One volunteer, anonymous in the study, said that though they made it to captain they found it stressful trying to stay ahead. So our options in reality two, where we need to resort to either hiding the task within a game or to using the kind of manipulation of reward that makes people feel like they're burdened with a task, are between projects which don't change people's thoughts about whether they can participate in science and those that feel like you've taken on a second job. In this reality, science is still reserved for the clever few who capture the attention and time of others through design, using the resulting effort for their own purposes. Participants are motivated not by curiosity, but by competition. It may be effective, but it seems a long way from the best that we could manage.

I want us to live in a third reality. This one is going to take some work, I think. It's a universe in which we don't need to rely on advances in machine learning to get the best out of the wealth of scientific data that we now have access to. It's one where human intuition and pattern recognition are still needed to get the most from data, even when machines are good at classification themselves. I feel pretty confident that this is indeed part of the reality we live in; though the recent advances in machine learning have been breathtaking, I think that our science is weird enough and our requirements exact enough that there will be a human element to it for a long while yet.

However, as I've repeatedly stressed, our ability to collect information about the Universe continues to astound. We shouldn't expect the pace at which new data flows in to decrease, nor should we expect it to become less open.* Because I want to keep communities of volunteers consciously participating in science, with no hiding in games, this means accepting that we won't be able to rely on citizen scientists to do all the work.

We've already opened the door to a solution. The supernova project showed that when humans and machines classify in concert, the outcome can be better than either working alone. I reckon that machines, as they improve while the surveys grow, will take on more of the burden, leaving the volunteers to review the unusual, the unexpected, and the interesting. The work comes in deciding how to divide the effort in such a way that allows the most interesting objects to be found; this probably means wading through a lot of confusing images or, for some projects, a lot of junk with little inherent interest. We need to understand how participants in Zooniverse projects want to work alongside the robot colleagues that my clever machine-learning colleagues are building for them.

This isn't a problem unique to us. In clicking your way around the material that Facebook chooses to show when you log in, you are in some sense collaborating with its algorithms. You are providing information about what the site should do next, which it responds to by showing you things it thinks you want to see.

* I'm somewhat dismayed that the LSST project has now taken money from international collaborators, including those of us in the UK, to help fund its operating costs in exchange for privileged access to data. I hope the leadership will see sense and, despite the need for cash to keep the lights on and the servers humming, find a way to go back to what was once imagined as the most open of projects, with data freely shared and available to everyone. The sky the telescope will scan, image, and monitor, after all, belongs to no one, and there is certainly plenty of science to go round.

More precisely, it will show you some combination of what it thinks you want to see, content that is most likely to expand the time you choose to spend on Facebook and content such as adverts that is profitable for the platform.

I hope that makes your skin crawl just a little. I think we're just beginning to understand how our attention is being manipulated on the internet, and to work out how to talk about it. I think that setting up a project like Galaxy Zoo, but with machine-learning classifiers actively working alongside human ones, is a fascinating problem which allows us to think about what we want. Even simple examples pose dilemmas. One worry is that if we allow a neural network to take the images it is most certain about away from classifiers then our poor humans might lose the brightest and most interesting images faster than the faint blobs, which are harder to classify. If we assume that people are, for all that they tell us they're in it for the science, partly interested in the brightest and most interesting galaxies, even if subconsciously the dopamine hit of suddenly seeing something spectacular keeps them classifying, then in allowing the machine to remove precisely these galaxies from circulation we have built a project which gets progressively less appealing over time.

Yet we have a problem. I can't just put the bright galaxies back in the pot without compromising on the promise, implicit in any citizen science project and explicit on the Zooniverse, that any work done by someone will actually be used. I experienced an early warning that this was going to be difficult when we shut down our original supernova project. As I said earlier, that decision was triggered when the researchers involved switched to an automatic classifier, and told me that they would no longer use the results of of citizen scientists' efforts.

On the face of it, an easy decision. The classifications were not being used, so we no longer had a project worth participating in.

Yet the participants, many of them dedicated people who would come back whenever new data was released, searching for supernovae time and time again, were not happy. I ended up calling a few of them to understand their views, and they were pretty unanimous in the fact they wanted their work to be used, for sure. No one I spoke to would participate in the project if we told them that we would just throw the results away. But they didn't understand why the researchers would switch to an automated system.

The researchers, I think, saw the automated system as just easier to understand. Given that a modern convolutional neural network can be essentially a black box, as inscrutable at least as the average person, I'm not sure that this is justified, but I see why they might think like that. We didn't think clearly enough about what this would feel like to the volunteers. One day they were contributing classifications that made a real difference scientifically. The next day they weren't, even though from their perspective nothing had changed; it's not as if they suddenly got worse. What had seemed to be a collaborative project was suddenly looking rather one-sided.

I think what we got wrong here was the lack of control we gave the volunteers, suddenly wrenching their project away from them, and I think that's the key to how to cope with the complexities of this third reality, when we combine human and machine classifications. If we give people control over what they see, they can make their own decisions about how they want things to run. I really like the idea of a project that says, 'We know if we give you more beautiful galaxies (or spectacular penguins, or interesting texts) you're likely to stay around longer. These classifications won't count, but do you want to see these images anyway? If so, how often?'

That seems honest and interesting, and I hope will lead to a system that can cope with the majority of the data heading our way.

If we don't do something like this, I worry that we'll miss out on the most unexpected of finds. On the contrary, I think we're likely—if we get it right—to be overwhelmed with interesting things.

Imagine a typical night, a few years from now when LSST is operating. As the Sun sets over the Chilean Andes, the dome containing the telescope opens up to allow it to cool in the cold night air. As the sky darkens, the enormous beast of a machine inside starts to methodically work its way across the sky, never pausing in one place for very long but often flicking back to where it has already been, to keep an eye out for asteroids and other rapidly moving or changing phenomena.

As the telescope and its camera work away, the images it takes are flowing digitally away from the mountaintop observatory and out into the world. They will soon end up at the US National Supercomputing Center in Illinois, where code will compare each one to previous images of the same field, checking the brightness of millions of objects. In any given image, millions of times a night, some object or other will be found to have changed in brightness, or to have apparently appeared from nowhere—or vanished completely.

LSST deals with this by issuing an alert, a public declaration of something happening in the sky. Its massive database, eventually laden with the fruits of ten years of surveying, will provide details of the history of each source. And then it's up to the rest of us. Software 'brokers' will try to filter this unprecedented torrent of data, sending the cosmologists pristine type 1a supernovae and planetary scientists a steadily growing list of candidate asteroids and Kuiper Belt Objects. One of these brokers will be listening too, but it will have a different job, directing objects to the screens of volunteers around the world.

Alerts will ping on a thousand mobile phones; something has happened in the sky, and we need your help. By the time the Sun rises in Chile, tens of thousands or maybe hundreds of thousands of images will have been inspected by a crowd consisting of both the astronomically passionate and the mildly curious. Sunrise in Chile means that it's nearly night in Australia, and we'll need to have identified the most interesting things by the time telescopes there are opening for the evening.

Maybe the centre of a nearby galaxy has brightened because something is falling into its black hole. Maybe a slow-moving object looks like a promising contender to be the latest member of the swarm of bodies out around Pluto. Maybe we caught a planet in transit in front of a star, or just the star itself behaving badly. Whatever the case, contributions from people like you will help determine what happens next. As the Earth spins, telescopes in Hawai'i and the Canary Islands and in South Africa join in; for the most energetic events, information from space-based satellites will be added to the mix.

For each of these events, triggering the worldwide network of observatories to stare at the right place is merely the start. Understanding what they are telling us will take a lot of time, and will overwhelm professional astronomers like myself. As data becomes more open, we'll see networks of citizen astronomers spring up to discuss and debate their favourite objects. Some of the participants are undoubtedly already experts in the field; some will bring skills that are of great use, and others just a willingness to learn. They will talk to and collaborate with the increasing number of scientists who have discovered just how powerful working in this way really is.

Between us, in this best of all possible worlds, we will have built a new way of exploring the Universe: something that takes the best features of Galaxy Zoo and Planet Hunters and all the other projects from the last decade and turns them into something even more inclusive, more powerful, and above all else much more fun, as volunteers control not only the discovery but the investigation of the things that are uncovered. I hope that if we look back in a decade's time at twenty years of citizen science through these projects, there will be a completely new crop of strange anomalies and curious objects to talk about. I would be very, very surprised to find myself in any reality other than this one.

It does, however, need you. You and your very human talent for pattern recognition and for being distracted from a task. You, with your curiosity and interest and willingness to spend just a few moments in your day doing something in contemplation of the Universe. You, with just a little time to join with millions of others so that collectively we can all achieve amazing things. Making the best of our capacity as a species to explore the Universe, and to understand the world around us, I believe, depends on finding a way that everyone on the planet can participate as an active observer and interpreter of the data that's now available. If we really can get everyone to join in—even if only for a few minutes—with this great endeavour, who knows what we might find, sitting out there and just waiting to be discovered.

# REFERENCES

I haven't tried to give a complete bibliography of works about citizen science, or even about the Zooniverse. An updated and nearly complete list of publications produced by Zooniverse projects is maintained at Zooniverse.org/publications, and the 'Meta' category there is a good starting point for those looking for academic studies of what we've been up to.

## PREFACE

Christiansen, Jessie L. et al., 2018, The K2–138 System: A Near-Resonant Chain of Five Sub-Neptune Planets Discovered by Citizen Scientists, *Astronomical Journal*, 155, 2, https://arxiv.org/abs/1801.03874.

Lakdawalla, Emily, 2009, An 'Amateur' Discovers Moons in Old Voyager Images, planetary.org, 5 August, http://www.planetary.org/blogs/emily-lakdawalla/2009/2035.html.

## CHAPTER 2

Bailey, Jeremy et al., 1998, Circular Polarization in Star-Formation Regions: Implications for Biomolecular Homochirality, *Science*, 281, 5377, 672.

Finkbeiner, Anne, 2010, *A Grand and Bold Thing*, Free Press. Provides a history of the Sloan Digital Sky Survey project.

Ivezić, Željko et al., 2018, *LSST: From Science Drivers to Reference Design and Anticipated Data Products*, https://arxiv.org/abs/0805.2366.

Reid, David and Chris Lintott, 1996, Astronomy at Torquay Boys' Grammar School, *Journal of the British Astronomical Association*, 5, 265, http://articles.adsabs.harvard.edu/full/1996JBAA..106..265R.

## CHAPTER 3

Land, Kate et al., 2008, Galaxy Zoo: The Large-Scale Spin Statistics of Spiral Galaxies in the Sloan Digital Sky Survey—Clockwise and Anticlockwise Galaxies, *Monthly Notices of the Royal Astronomical Society*, 388, 1686.

Lintott, Chris et al., 2008, Galaxy Zoo: Morphologies Derived from Visual Inspection of Galaxies from the Sloan Digital Sky Survey, *Monthly Notices of the Royal Astronomical Society*, 389, 1179, https://arxiv.org/abs/0804.4483.

Lintott, Chris, Ignacio Ferreras, and Ofer Lahav, 2006, Massive Elliptical Galaxies from Cores to Halos, *Astrophysical Journal*, 648, 826, https://arxiv.org/abs/astro-ph/0512175. This is the work I was presenting in Oxford in my job interview.

Schawinski, Kevin et al., 2007, Observational Evidence for AGN Feedback in Early-Type Galaxies, *Monthly Notices of the Royal Astronomical Society*, 382, 1415, https://arxiv.org/abs/0709.3015. Kevin's 50,000 galaxies.

Schawinski, Kevin et al., 2009, Destruction of Molecular Gas Reservoirs in Early-Type Galaxies by Active Galactic Nucleus Feedback, *Astrophysical Journal*, 690, 1672, https://arxiv.org/abs/0809.1096. Observations from IRAM.

## CHAPTER 4

Andersen, Katharine, 2005, *Predicting the Weather: Victorians and the Science of Meteorology*, University of Chicago Press.

Argelander, Friedrich, 1844, Aufforderung am Freunde der Astronmie zur Amstellung von ebenso interssamten und nützlichen, als leicht auszuführenden Beobachüngen über mehrer wichtige Zweige der Himmelskunde, in *Astronomisches Jahrbuch*, ed. Schumacher.

Darwin's correspondence is brilliantly presented at https://www.darwinproject.ac.uk/.

Mantell's letter is Letter no. 1663, accessed on 22 November 2018, http://www.darwinproject.ac.uk/DCP-LETT-1663.

Pasacoff, Jay, 1999, Halley as an Eclipse Pioneer: His Maps and Observations of the Total Solar Eclipses of 1715 and 1724, *Journal of Astronomical History and Heritage*, 2, 39.

Robin's appeal is in the *Gentleman's Magazine*, December 1748, Rules for Computing the Height of Rockets, &c, 18, 597 and his one correspondent's letter is in the May 1749 *Gentleman's magazine*, 19, 217.

Rowson, Ben and William O. C. Symondson, 2008, Selenochlamys ysbryda sp. nov. from Wales, UK: A Testacella-like Slug New to Western Europe (Stylommatophora: Trigonochlamydidae). *Journal of Conchology*, 39, 5, 537–52.

Shuttleworth, Sally, 2015, Old Weather: Citizen Scientists in the 19th and 21st Centuries, *Science Museum Group Journal*, 10, 15180, 150304.

## CHAPTER 5

Barnard, Luke et al., 2015, Differences Between the CME Fronts Tracked By an Expert, an Automated Algorithm, and the Solar Stormwatch Project, *Solar Physics*, 13, 709.

Bugiolacchi, Roberto et al., 2016, The Moon Zoo Citizen Science Project: Preliminary Results for the Apollo 17 Landing Site, *Icarus*, 271, 30, https://arxiv.org/abs/1602.01664.

Eveleigh, Alexandra et al., 2013, 'I want to be a Captain! I want to be a Captain!': Gamification in the Old Weather Citizen Science Project, in *Gamification '13: Proceedings of the First International Conference on Gameful Design, Research, and Applications*, ACM, New York.

Jones, Shannon et al., 2017, Tracking CMEs Using Data from the Solar Stormwatch Project: Observing Deflections and Other Properties, *Space Weather*, 15, 1125.

Mutch, Simon, Darren Croton, and Gregory Poole, 2011, The Mid-Life Crisis of the Milky Way and M31, *Astrophysical Journal*, 736, 84.

Simmons, Brooke, Rebecca Smethurst, and Chris Lintott, 2017, Supermassive Black Holes in Disc-Dominated Galaxies Outgrow Their Bulges and Co-evolve with Their Host Galaxies, *Monthly Notices of the Royal Astronomical Society*, 470, 1559.

Smethurst, Rebecca et al., 2017, Galaxy Zoo: The Interplay of Quenching Mechanisms in the Group Environment, *Monthly Notices of the Royal Astronomical Society*, 469, 3670.

van der Marel, Roeland et al., 2012, The M31 Velocity Vector. III. Future Milky Way M31–M33 Orbital Evolution, Merging, and Fate of the Sun, *Astrophysical Journal*, 753, 9.

Watson, David and Luciano Floridi, 2018, Crowdsourced Science: Socio-technical Epistemology in the E-research Paradigm, *Synthese*, 195, 2, 741.

## CHAPTER 6

Charcot, J.B., 1980, *The Voyage of the 'Why Not?': The Journal of the Second French South Polar Expedition, 1908–1910*. Translation by Philip Walsh. University of Toronto Library, The Musson Book Company Limited.

Jones, Fiona et al., 2018. Time-Lapse Imagery and Volunteer Classifications from the Zooniverse Penguin Watch Project, *Scientific Data*, 5, 180124.

Meyer-Rochow, Victor Benno, and Jozsef Gal, 2003, Pressures Produced When Penguins Pooh—Calculations on Avian Defaecation, *Polar Biology*, 27, 56.

## CHAPTER 7

Bowyer, Alex et al., 2015, This Image Left Intentionally Blank: Mundane Images Increase Citizen Science Participation, *Human Computation*, https://www.humancomputation.com/2015/papers/45_Paper.pdf.

Fossey, Steve et al., 2014, Supernova 2014J in M82 = Psn J09554214+6940260, *Central Bureau for Astronomical Telegrams*, 3792.

Geach, J.E. et al., 2015, The Red Radio Ring: A Gravitationally Lensed Hyperluminous Infrared Radio Galaxy at z = 2.553 Discovered through the Citizen Science Project SPACE WARPS, *Monthly Notices of the Royal Astronomical Society*, 452, 502, https://arxiv.org/abs/1503.05824.

Marshall, Philip et al., 2016, SPACE WARPS—I. Crowdsourcing the Discovery of Gravitational Lenses, *Monthly Notices of the Royal Astronomical Society*, 455,1171, https://arxiv.org/abs/1504.06148.

Smith, Arfon et al., 2011, Galaxy Zoo Supernovae, *Monthly Notices of the Royal Astronomical Society*, 412, 1309, https://arxiv.org/abs/1011.2199.

Swanson, Alexandra et al., 2015, Snapshot Serengeti, High-Frequency Annotated Camera Trap Images of 40 Mammalian Species in an African Savanna, *Scientific Data*, 2, 150026.

Wright, Daryll et al., 2017, A transient search using combined human and machine classifications, *Monthly Notices of the Royal Astronomical Society*, 472, 1315, https://arxiv.org/abs/1707.05223.Chapter 8

Banfield, Julie et al., 2015, Radio Galaxy Zoo: Host Galaxies and Radio Morphologies Derived from Visual Inspection, *Monthly Notices of the Royal Astronomical Society*, 453, 2326, https://arxiv.org/abs/1507.07272.

Cardamone, Carolin et al., 2009, Galaxy Zoo Green Peas: Discovery of a Class of Compact Extremely Star-forming Galaxies, *Monthly Notices of the Royal Astronomical Society*, 399, 1191, https://arxiv.org/abs/0907.4155.

Chabris, Christopher et al., 2011, You Do Not Talk About Fight Club if You Do Not Notice Fight Club, *i-Perception*, 2, 150.

Drew, Trafton, Melissa Vo, and Jeremy Wolfe, 2013, The Invisible Gorilla Strikes Again, *Psychological Science*, 24, 1848.

Józsa, G.I.G. et al., 2009, Revealing Hanny's Voorwerp: Radio Observations of IC 2497, *Astronomy & Astrophysics*, 500, 33, https://arxiv.org/abs/0905.1851.

Lintott, Chris et al., 2009, Galaxy Zoo: 'Hanny's Voorwerp', a Quasar Light Echo? *Monthly Notices of the Royal Astronomical Society*, 399, 129.

## CHAPTER 9

Boyajian, Tabetha et al., 2016, Planet Hunters IX. KIC 8462852—Where's the Flux? *Monthly Notices of the Royal Astronomical Society*, 457, 3988, https://arxiv.org/abs/1509.03622.

Boyajian, Tabetha et al., 2018, The First Post-Kepler Brightness Dips of KIC 8462852, *Astrophysical Journal*, 853, 8, https://arxiv.org/abs/1801.00732.

OED Online. July 2018. 'white lie, n.' Oxford University Press. http://www.oed.com/view/Entry/420925?redirectedFrom=white+lie& (accessed November 26, 2018).

Wright, Jason et al., 2018, The Search for Extraterrestrial Civilizations with Large Energy Supplies. IV. The Signatures and Information Content of Transiting Megastructures, *Astrophysical Journal*, 816, 17, https://arxiv.org/abs/1510.04606.

## CHAPTER 10

Gal, Yarin and Zoubin Ghahramani, 2015, Bayesian Convolutional Neural Networks with Bernoulli Approximate Variational Inference, http://arxiv.org/abs/1506.02158.

Kim, Jinesop et al., 2014, Space-Time Wiring Specificity Supports Direction Selectivity in the Retina, *Nature*, 509, 331.

Sullivan, Devin et al., 2018, Deep Learning is Combined with Massive-Scale Citizen Science to Improve Large-Scale Image Classification, *Nature Biotechnology*, 36, 820.

Walmsley, Mike et al., 2019, Identification of Low Surface Brightness Tidal Features in Galaxies Using Convolutional Neural Networks, *Monthly Notices of the Royal Astronomical Society*, 483, 2968.

# LIST OF FIGURE CREDITS

1 NASA/JPL-Caltech
2 Courtesy of Ted Stryk
3 Nicolas Kipourax Paquet/Getty Images
4 Fermilab Visual Media Services
7 galaxyzoo.org
8 Sloan Digital Sky Survey
9 Sloan Digital Sky Survey
10 Pictorial Press Ltd/Alamy Stock Photo
11 Karen Masters and SDSS
12 © DiVertiCimes
13 DEA/G DAGLI ORTI/De Agostini Editore/age fotostock
14 Royal Astronomical Society/Science Photo Library
15 Biblioteca Ambrosiana/De Agostini Editore/age fotostock
16 Mantell Family Collection, Alexander Turnbull Library, Wellington, New Zealand (MS-Papers-0083-268-003)
17 NASA, ESA, B. Simmons (Lancaster University/UC San Diego) and J. Shanahan (UC San Diego)
18 NASA's Goddard Space Flight Center/ASU
19 The National Archives
20 Photo by Chris Lintott
21 Photo by Chris Lintott
22 Snapshot Serengeti
23 Trafton Drew et al: 'The Invisible Gorilla Strikes Again: Sustained Inattentional Blindness in Expert Observers', *Psychological Science* 24 (9), 1848–53, 2013. 10.1177/0956797613479386.
24 William C. Keel et al., 'HST Imaging of Fading AGN Candidates I: Host-Galaxy Properties and Origin of the Extended Gas', Published 2015 April 14 • © 2015. The American Astronomical Society. All rights reserved.
25 Carolin Cardamone et al., 'Galaxy Zoo Green Peas: Discovery of a Class of Compact Extremely Star-Forming Galaxies', *Monthly Notices of the Royal Astronomical Society*, Volume 399, Issue 3, 1 November 2009, Pages 1191–205, https://doi.org/10.1111/j.1365-2966.2009.15383.x © 2018 Oxford University Press
26 H. Giguere, M. Giguere/Yale University

27 T. S. Boyajian et al., 'Planet Hunters IX. KIC 8462852: Where's the Flux?', *Monthly Notices of the Royal Astronomical Society*, Volume 457, Issue 4, 21 April 2016, Pages 3988–4004, https://doi.org/10.1093/mnras/stw218. © 2018 Oxford University Press

28 Mark Garlick/Science Photo Library

# LIST OF PLATE CREDITS

1 David St. Louis (CC BY 2.0)
2 M. Blanton and SDSS
3 ALMA (ESO/NAOJ/NRAO)
4 ESO
5 (a) NASA, ESA, S. Beckwith (STScI), and The Hubble Heritage Team (STScI/AURA); (b) NASA, ESA, and The Hubble Heritage Team (STScI/AURA)
6 NASA, H. Ford (JHU), G. Illingworth (UCSC/LO), M. Clampin (STScI), G. Hartig (STScI), the ACS Science Team, and ESA
7 RAL Space HI instrument team, the STEREO mission and SECCHI teams, and NASA
8 Tommy Eliassen/Science Photo Library
9 Adam Block/Mount Lemmon SkyCenter/University of Arizona
10 J. E. Geach et al., 'The Red Radio Ring: A Gravitationally Lensed Hyperluminous Infrared Radio Galaxy at z = 2.553 Discovered through the Citizen Science Project SPACE WARPS', *Monthly Notices of the Royal Astronomical Society*, Volume 452, Issue 1, 1 September 2015, Pages 502–10. © 2018 Oxford University Press
11 NASA, ESA, William Keel (University of Alabama, Tuscaloosa), and the Galaxy Zoo team

# INDEX

*Note*: Figures and footnotes are indicated by an italic *f* and *n*

**R**



**S**

# THE COSMIC MYSTERY TOUR

Nicholas Mee

978-0-19-883186-0| Hardback | £16.99

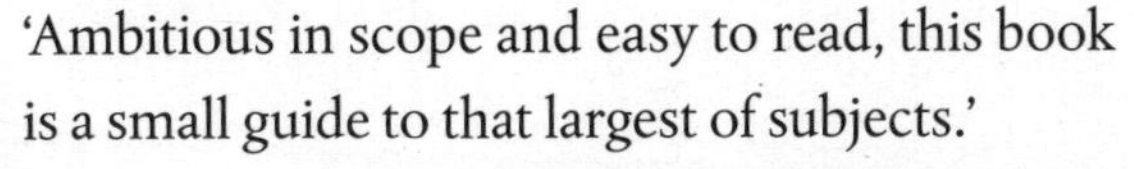

'Ambitious in scope and easy to read, this book is a small guide to that largest of subjects.'
Pippa Goldschmidt, *BBC Sky at Night*

'An accessible introduction to all things cosmos…' Maren Ostergard, *Booklist*

*The Cosmic Mystery Tour* takes us on a lightning tour of the mysteries of the universe enlivened by brief stories of the colourful characters who created modern science. It explores a range of hot topics in physics and astronomy, including the recent discovery of gravitational waves, the quest for the origin of dark matter, and the study of the supermassive black hole at the centre of the galaxy.

This lively, inspiring book is the perfect beginner's guide to how the universe works.

# QUANTUM SPACE

*Loop Quantum Gravity and the Search for the Structure of Space, Time, and the Universe*

Jim Baggott

978-0-19-880911-1 | Hardback | £20.00

'*Quantum Space* is the first complete and approachable account to a quantum theory that delves deep into the quest to resolve one of the great unanswered problems of modern physics.'

*All About Space Magazine*

'The discovery and development of Loop Quantum Gravity has been a great adventure. Jim's story beautifully captures its insights and excitement.'

Carlo Rovelli, author of *Seven Brief Lessons on Physics*

A fascinating discussion of the quest to resolve one of the great unanswered problems of modern physics: how can general relativity and quantum mechanics be compatible? Jim Baggott tells the story through the careers and pioneering work of two of the theory's most prominent contributors, Lee Smolin and Carlo Rovelli. Combining clear discussions of both quantum theory and general relativity, this accessible book offers one of the first efforts to explain the new quantum theory of space and time.

# THE ETHICAL ALGORITHM

*The Science of Socially Aware Algorithm Design*

Michael Kearns and Aaron Roth

978-0-19-094820-7 | Hardback | £18.99

Understanding and improving the science behind the algorithms that run our lives is quickly becoming one of the most pressing issues of this century; traditional solutions, such as laws, regulations and watchdog groups, have proven woefully inadequate at best. Derived from the cutting-edge of scientific research, *The Ethical Algorithm* offers a new approach: a set of principled solutions based on the emerging and exciting science of socially aware algorithm design.

In this new release, Michael Kearns and Aaron Roth explain how we can better embed human principles into machine code without halting the advance of data-driven scientific exploration. Weaving together innovative research with stories of citizens, scientists, and activists on the front lines, *The Ethical Algorithm* offers a compelling vision for a future, one in which we can better protect humans from the unintended impacts of algorithms while continuing to inspire wondrous advances in technology.

# THE WOMEN OF THE MOON

*Tales of Science, Love, Sorrow and Courage*

Daniel R. Altschuler and Fernando J. Ballesteros

978-0-19-884441-9 | Hardback | £20.00

Although the great poet Dante described the Moon as being 'smoothly polished, like a diamond', philosophers in times past acknowledged that its appearance was in fact extremely uneven, and sought to understand why. The agreed conclusion was that, like a mirror, it reflected an imperfect Earth.

The Moon, or more precisely the nomenclature of lunar craters, still holds up a mirror to an important aspect of human history: of the 1586 craters named to honour philosophers and scientists, only twenty-eight honour women. Not only do these women of the Moon present us with an opportunity to meditate on this gap, but perhaps more significantly, they offer us an opportunity to talk about their lives, mostly unknown until now.

# ANTIMATTER

*Second Edition*

Frank Close

978-0-19-883191-4 | Oxford Landmark Science | Paperback | £8.99

'Beautifully written...This book will inspire a sense of awe in even the most seasoned readers of physics books.'

Amanda Gefter, *New Scientist*

'Beautifully concise history of one vital aspect of twentieth-century particle physics.'

Mark Ronan, *Times Literary Supplement*

*Antimatter* explores a strange mirror world, where particles have identical yet opposite properties to those that make up the familiar matter we encounter every day; where left becomes right, positive becomes negative; and where, should matter and antimatter meet, the two annihilate in a blinding flash of energy that makes even thermonuclear explosions look feeble by comparison. Frank Close separates the facts from the fiction about antimatter, and explains how its existence can give us profound clues about the origins and structure of the universe.

Oxford Landmark Science books are 'must-read' classics of modern science writing which have crystallized big ideas, and shaped the way we think.

ANDER, MEGAN A JOY, RUTH SEYMOUR, HUY TRAN, SUNGJUNC@TASHSCHOOL.ORG, LISA
OLEIN, KAY ROZEBOOM, LOLO, BATIST, NATH, LUCY HEREFORD, QUYNH TRAN, BRIANK, KRIS
HERYL ROBBINS, ROXANNE M TRIPP, ROBIN JOHNSON, SMWEAVE, ANDREW THOROGOOD, K
LEE, OREGANO, WIM TEMMERMAN, PETER GREINER, ELERI HUGHES, KELLY BINNINGTON,
DREK, STEVE185@UMN.EDU, NICOLE ABANTO, ATOMICA, STEVEN READ, ETHAN HAWKINS, G
R, MASON SLADE, 2FAY, SASKIA CRAMER, OSMAN MICHAEL JIRON, FLORIAN BLAESS, SUE
EAR, PENNY A. PETERSEN, BRIAN WICKS, OLFAR, MAIRE, LINDSAY ZAREMSKI, JACI JACKSON
CHOLZIA, JOE LAPORTE, KAT STOKES, L.A. VAN DEN HEUVEL, GARRY PENNELL, BRENT BEA
BOFSKY, SIONED PEAKE, PSPENCE, REED SLOAN, WESTELL BEECHEY, ZUZANA MACH√ ČKO
CHETT, JESSICA MANZAK, GEOFF ROBERTS, IR78, CHERIS J LIFFORD, MAJOSK, GABRIELA S
LIZZIE MEADE, COSMOSLACKIST, MOONMARE, CAROL ANN SMITH, KATHBAILEY, GRACIE
ABOB, BARBLEERN, SAIGESIMCOX89, DANIEL, ELIZABETH MCLENDON, ROBYN DELES, JOSHUA
ION, SUSAN PRESLEY, RUGENIUS, JAKOB BURCHARD, STONE, LIZDENNY, ROBERT G LAWL
ENSIR, DANIEL MARQUARDT, LYSS, ALLAN, BOONYARAT, FRANK KICK, WILLEM FORREST, DE
ZEKE LAFROMBOISE, TALIA TOLBERT, VERNON SOUSA, SUZANNE STRAPP, ANTONYK, CARO
N, ELEANOR MCVEY, MARIO MARANI, CHRISTOPHER WOLLMAN, KATHY OLMSTED, LINDA
ENIC WILEY, PHILIPPE ZIGRAND, ADAM A MILLER, NFLUMERF@UWO.CA, ALEC HOLDEN, RO
AS STARNES, WILLIAM P HALL, MIKEYATES, GERARDO, DEBORAH ATTARIAN, ALI AHMED, R
ROBERT H PODOLSKY, KARRI NEUFELDT, HUNTER TIBOR, OSCARMENTAL, LUCY BRANFORD
XPLORA, CHRIS YI, CEBALDWIN, SAM, ANDYTRON, PONNUKI, KALE KRIENKE, BRENT OGBUR
MEYERS, AVV, JARED WHITE, CHIKELIGHTENING, SHIRLEY CHIRCH, MARCO BONAFINI, SAMIR
ACHO, GURUGURU, ADAM CLIMIE, JO BLACKMAN, DALYN BANAGAS, ISABEL, RYAN DARNELL, I
R, CANDICE, ANGELICA, BECKI BISHOP, SKUZA005, HOWARD, CHRISTIAN PHILLIPS, SAGE AN
MITH, HARRY E. BEDDER, HTSUDA, KIM REYNOLDS, TRAVIS ROLINE, EMILY WELSH, DOCJUKI
H, PAWE≈C SZA≈ČEK, RAY, PLENUM, PAVEL VALKANOVSKY, CHRISDUNSEATH, VINCENZO LIZ
E, LAURATANJI, BETH WHITELEY, ASTRO1005, MATTHEW GORMAN, ALAN BELL, CARALIBRA, TU
DO, KTODD, CHRISTINE PREIMESBERGER, LAURAJOHNSON, DEANNA NORMAN, OCTODRAGO
O, CALIFORNIANCO6@GMAIL.COM, JACOB, KAI WACKER, BARBARA ROBERTS, MASTERCOMPOS
LYNKAPAS, HEIDI MOSS, GEOFFREY LEE, EDMISTJA@GMAIL.COM, STEVEN LOFTHOUSE, LI
HY, CELTICAIRE, TINKAPUPPY, WQFEWEQ WQFEFQFE, 7356445735Y, ZO√ VARLEY, VICKI MC
K BLEASE, CZARRE45, ELIZABETH BROPHEY, CHENAY SIMMS, EVAN BASS, DARION HARRIS,
NGHAUSEN, PAPERRY, MOMMAJOAN, JASON BRANDLER, RICK, BRODY, MINDSTATE56, THEFA
JANNJORD, CHELSEY MAURER, SRMCLEAN124, RANDALL MICHAL, JLIVEWELL, RYAN ROSS,
IWA, B L GOODWIN, RUTH M. POTRAFKA, STEVEN & ANDREW BLODGETT, KRYSTLE, DIANE R
WILLOWSKYE, ORTWIN RAU, REBECCA LAWSON, DIANA P. MICHELLE, C MASON, ASTRO
BAYEAGLE91, KATHRYN SCHWADER, ALYSSA BRAULT, IANRYAN6@BIGPOND.COM, LOUISE E
EVIE STEVENS, NZNOTINAS, KATH MULLHOLAND, GRAHAM JONES, L8ENUF, GUILLE, KRISTIN
OTTER, CARYN JEFFERY, OMELDA ARISTIL, AQUADOMUS, JOHN HUTTER, HELEN BENNETT,
YO, MARGARET LEWIS, ANNIEPSOUTHWEST, TAFADZWA TAZVITADZA, M57RING, MATTHEW
ER, YAMMIR RAMOS, ZEUSMOOSE, FEDNUIK3, MATT SIMPSON, ALLAN BROWN, OCTAVE MUL
IONA, PHOEBE BRACKEN, STORM, ANDR√©S GUILLERMO STENNER, VXUMA, JIMMY LAMB,
STEKI, OSCAR, SANDY, ARWINGL52, VESQ, BARRYWSTUART, AEJSS5, HERBERT SCHLETTER,
ER, CONNOE GRANT, HILOHELLO, MARCO TISON, ALLYSON MCDOWELL, JEISAS, MICHAEL W
CHANCE CHAMPAGNE, KATRINA BARAUSKAS, POPA, FELIPE ROLIM TRISTAO, PEPE J HERNA
ZIADYK, JARED AGNEW, 21OPPGAV@STU.PLATTEVILLE.K12.WI.US, BARNES, CLIFF, LANDON AL
O.COM, ALEX SANG, MARTIN COCKETT, MEGALONYX, GARDENMAEVE, COLL7723, COLCOL
HI, FED6, MAKSYM ABRAMOV, MARCEL ROTHEN, CHRISTIE STRINGER, BEATRIZ AGUILAR BAR
TOPHER GRIPP, PAVEL GAVRILENKO, AMILY, MAX BARNETT, GORDON FLOWER, JUSTIN SCH
MAMAGOTCHA, LAURA M. EDWARDS, PATRICK SMITH, ERIK JENSEN, KRISTOFER KIRKPATRI
N, BEN HOLLINGBERY, PHILJPWADE, MARIA PETZOLD, TRANG DO, KEVIN UNDERWOOD, C
NSKI, JANET SBFINATI, MATTHEW M, JACKSTOGO, CKAMPRATH, NOTORIOUSVHD, JOHN SEE
BOONE, MRSALPHAGEEK, EMGONNUSCIO, KRISTIEN, NICOLE STENBECK, JORGE A. CAZAR
N CARDEN, DIANE WRIGHT, BEN, BROOKE SAYRE-CHAVEZ, BENJAMIN VON ESCHEN, MICHAE
DAVID JACOBY, ALBERT RICHTER, CANIS778, PRSPT@AOL.COM, THOMAS MITCHELL, TERANE
RIEND, VANRENTER, ANTONY BADUNS, ANTONETTI, JAMES RABE, JONATHAN N. LEYLAND, B
APHACH SRIKONGPHLEE, DONG YUAN YANG, PAIGESCHLENKER, CYNTHIA F SEEBAUER, TAM
TUS, ALISON BOYD, SUZZGRANTT, JAN LAMBERT, VICENTE MINGUEZ, SALLYE, BRIAN HERTZIG
VANCOR, TEAGAN WILLIAMS, PIERRE GONNET, ANDREW SOTO, HANNAH THOMAS, STEFAN
BERUBE, JORDAN TAYLOR, HOPE STEVENS, KERRY KELLY, JOANNE HOPKIN, DUSTIN, CATN
HEADSMUM, GHADY, ISAK WHITE, DONNA, MARY KATHERINE CARNES, GRIFFIN ROCK, HELEN
ILLIE, EMITA, REBECCA DONOGHUE, ELIZABETH FOX, ANEKA SWANSON, SARAH ELIZABET
MARY A. NYGREN, CHARLOTTEISCOOL, DOUG GOLDSMITH, KATELYN CAFARELLI, CAITLYN, M
IKEYAN, D. FHEN, THOMAS, SHELBY GARZA, FSBRENNAN, THOMASLOFTEN, AIDAN FLANNE
ELIZABETH DANZIG, CHARLIE JENNINGS, SIMONE AIELLO, SIDDHARTHJOHARY@GMAIL.CO
SYLVIA HERREN, MICHELLE MATTESON, DALE BORGESON, ETHAN THOMAS, GENTOOMICH
GOV, BARBARA GORDON, JESSICA GILLIAM, TOM GALBRAITH, JEFFREYHRWTZ, PHILIP HER
ALEXANDER G SANDSTROM, LECHITO SEGURA, WILL STEIN, ROBERTLEWIS2, SERCOUTREAC
MARCINCSAK, MARIO AZV, CAROL POPE, IAIN DOWNIE, ATHBIRDIE, MARY A. READ, GIANN
LSON, DAMIEN BRICARD, RAULE, TERNGIRL, LINDA MOODIE, PAUL IBBETSON, ZIAD KAWASH
ICHOLSON, MIKE CHAFFEE, PKP_1232001, AARON WALDEN, ARTUR RACZYNSKI, MARTHA
ATHY, TAMI LOCKARD, ADELAIDE BURGER, PAULINE ENGUEHARD, MARCO FONSECA, BAR
DO 100, NAOMI, ABROUGHA, ZORROBIS, MARIANNE ARMSTRONG, DICK CHENEY, DLPARTN
STAD, GEEK3, MADELINE CLANCY, LG PRICE, SUJAY VAIDYA, ELWEBSTER, CAROL E SHYA
R, DAFF43, FOREST, KHAYRIAH, COUTAIN, MORALE DANIEL, BOBBY VANDENBUSH, RYBE
UCILE, SANJAY MEHTA, BONNIEC, KATHERINE MONTEIRO, MARCUSBLOCK, VCJOHNSON, C
IBEDOSD1@SVSABERS.ORG, JOHN FUNCHES, JUDAICA DH AT THE PENN LIBRARIES, CAROL
OLITOR, CHARINPINES, HISTORYNERD1109, KELSI SNIDER, DAVE CARTER, KENTUCKGAL, RA
ON, LIV KANE, ESTRELLA GUTI√©RREZ, AFTAB ALI, KELLI APPELEN, NICK, CONNIE MCADAM
SVENDSEN, JOSHUA, PAULINE WHITING, JENNY H, MARCO SOMMACAMPAGNA, GEORGE PE
OLSEN, SGRAY8145, JAMES C. MERRITT, DICK M, MARIE FAUCHER, PANXFONEM, H3LLY, M
LYDEAH, HEPBURN, SCOTT THOMPSON, SKORBOLAMID, CAROLYN GREENE, TROPIC LUCA